An Introduction to Medicinal Botany

The Authors

Dr. Showket Ahmad Ganaie (M.Phil, Ph.D, M.Ed) is presently working as a Teacher in the Department of School Education, Govt. of Jammu & Kashmir. Dr. Ganaie worked as Lecturer in Botany (Contractual) in the Department of Higher Education, Govt. of Jammu & Kashmir from 2016 to 2019. He has done extensive work on Pharmacognosy and Phytochemistry of diverse medicinal plant species. He has published 46 research papers in various international and national journals of repute and he has presented in more than 25 scientific conferences/seminars and attended several workshops & training courses on different aspects of biology.

Dr. Surender Singh Yadav (M.Sc, CSIR NET, Ph.D, PDF) is presently working as Associate Professor in Department of Botany, Maharshi Dayanand University Rohtak. Dr. Yadav did extensive work in the field of Ecology, Environment & Biodiversity. He has published 70 papers in national and international journals of repute, authored/edited several books & magazines and guided 07 candidates for Ph.D degree. He has received University Gold Medal in quiz competition, (NSS) Merit Certificate, First position in National Level Debate Competition & Certificate of Appreciation by Department of Forest, Govt. of Haryana.

Dr. Rohi Jan (M.Phil, Ph.D, M.Ed) has pursued her Doctorate in Chemistry from Savitribai Phule Pune University. Dr. Rohi worked as Research Associate awarded by CSIR, New Delhi. She has expertise in the field of Environmental Toxicology. Dr. Rohi has published 15 research papers in various international and national journals with high impact factors and attended various scientific conferences, seminars, workshops & training courses.

Dr. Waseem Raja (M.Sc, Ph.D) is currently working as Assistant Professor (Botany) in the Department of Higher Education, Govt. of Jammu & Kashmir. He has published a good no. of research papers in the international and national journals.

Dr. Mohd Shahijahan Dar (M.Sc, Ph.D) is presently working as Senior Assistant Professor in Botany in the Department of Higher Education, Govt. of Jammu & Kashmir. He has expertise in the field of Plant Pathology and has published various research papers in international & national journals.

An Introduction to Medicinal Botany

Showket Ahmad Ganaie
Department of School Education
Govt. of Jammu & Kashmir, India
(M.Sc., M.Phil., Ph.D., B.Ed., M.Ed.)

Surender Singh Yadav
Assistant Professor
Department of Botany
Maharshi Dayanand University Rohtak (Haryana), India
(M.Sc., Ph.D., PDF)

Rohi Jan
Research Associate
Department of Chemistry
Savitribai Phule Pune University, India
(M.Sc., M.Phil., Ph.D., B.Ed., M.Ed.)

Waseem Raja
Assistant Professor
Department of Botany
Govt. Degree College Pulwama, Jammu & Kashmir, India
(M.Sc., Ph.D.)

Mohd Shahijahan Dar
Assistant Professor
Department of Botany
Govt. Degree College Pulwama, Jammu & Kashmir, India
(M.Sc., Ph.D.)

2025

Daya Publishing House®

A Division of

Astral International Pvt. Ltd.

New Delhi – 110 002

ISBN: 978-93-5919-998-6

Published by : **Daya Publishing House®**
A Division of
Astral International Pvt. Ltd.
– ISO 9001:2015 Certified Company –
4736/23, Ansari Road, Darya Ganj
New DelhiA110 002
Ph. 011A43549197, 23278134
EAmail: info@astralint.com
Website: www.astralint.com

Preface

Medicinal Botany is an emerging and fast growing branch of biology with multidimensional approach. Currently various reputed scientific and academic institutions are engaged in doing research work in the different disciplines of medicinal botany, including ethnobotany, ethnomediicne, pharmacognosy, pharmacology, phytochemistry, pharmaceutics, phytomedicine and other related fields. Looking to its scope and future prospects, many universities in India have started teaching programmes on medicinal botany at undergraduate and post graduate level. In many modern laboratories, reseerachers are working on molecular aspects of medicinal plants. Throughout the globe, developing countries in particular, the subject of medicinal botany has gained popularity. Researchers, scholars, academecians need a handy book giving essential information on basic concepts, definition, approaches and applictions in such subject.

Authors are pleased to place before the readers, the first edition of the book "Introduction to Medicinal Botany". The present book is an attempt to provide with the fundamentals of various concepts in medicinal botany especially keeping into consideration the needs of the students as per the choice based credit system curriculum introduced by University Grants Commission (UGC). This book has been designed and written in a manner to meet the aspirations of undergraduate and post graduate students and research folk working in the field of herbal medicine, pharmacognosy, analytical pharmacognosy, pharmacology and phytochemistry. The significant feature of this book is its clear, accurate and up-to-date information. We have given equal importance to text, figures and tables for textual clarifications. It is written in clear and concise language to enhance self-motivation.

Suggestions for the improvement of the book will be thankfully acknowledged and incorporated in further editions.

Authors

Contents

UNIT 1

Herbal Medicine

1.1 Introduction

Herbal medicines (Herbalism), also called Botanical medicines or Phytomedicines, are those medicinal products which are prepared by using mainly herbs or plants, plant parts and plant products with minimal or no industrial processing or chemical manipulations, used for medical and therapeutic purposes.

Herbal medicines have long earned reputation as "the people's medicines" because of their easy accessibility, safety, and the ease with which they can be prepared. Use of herbal medicines in the treatment of diseases has a long tradition of use outside of conventional medicines. This method is now gradually scorning a more mainstream method of treatment in many countries of the world, because improvements in the methods of analysis and quality control of herbs and herbal drugs along with advances in clinical research and their efficacy and safety are continuously bringing light to the value of herbal medicines in the prevention and treatment of diseases. In some Asian and African countries, 80 per cent of the population depends on traditional herbal medicines for primary Healthcare. In many developed countries, 70-80% of the population have used some form of complementary or alternative medicines (CAM) composed primarily of herbal medicines.

Thus, in a broader sense, it may be said that "Herbal Medicines" are the finished herbal medicinal products, which contain not only herbs, but also herbal products like naturally occurring plant-derived substances such as fresh juice, latices, extracts, gums, fixed oils, essential oils, resins, tannins and dry powders of herbs, which may be processed by various local procedures, such as steaming, boiling with water, roasting, or stir-baking with honey, alcoholic beverages or other materials and herbal preparations such as powdered herbal materials, or their extracts, tinctures, and fatty oils as active ingredients and are used in the treatment of illness and for maintenance of health and well-being. However, the scope of herbal medicines is sometimes extended to include fungal and bee products, as well as smaller amount of minerals, shells and certain animal parts. Finished herbal products may also contain excipients in addition to the active ingredients.

1.1.1 Benefits and advantages of Herbal Medicines

The indications of herbal medicine are diverse, being employed for the treatment of a wide range of diseases, when compared to conventional medicines. Herbal

Medicine remains a major component of the primary healthcare in many rural communities. It constitutes an integral part of the culture of many societies of the world. Many herbs and herbal recipes have a long traditional history of folk uses and claims of health benefits. Herbal medicines contain complex chemical compounds that are responsible for the pharmacological activities, which correspond to health benefits.

Safety and efficacy is an important factor for the usage and commercialization of herbal medicines. The quality of herbal products is largely dependent on the safety and efficacy of the herbal material.

In contrast to the synthetic pharmaceutical drugs, which often consists single chemical entity, the herbal medicines are typically made up of numerous compounds usually in the crude, unpurified state. Many finished herbal products are made from folk recipes often containing more than one herbal material as the active component. The composite nature of most herbal drugs may be the reason for many of their benefits. The multicomponent ingredients may boost benefits by enhancing simultaneously certain important pharmacological activities. In addition, some constituents may also act on more than one physiological system, thereby confers wide range of therapeutic benefits to it. The major advantages of herbal medicine are:

- They are affordable (cheaper)
- Reduced risk of side effects
- There is a choice on how to use them
- They are good for more than one condition (multipurpose)
- Effectives with chronic conditions
- Widespread availability
- Increasingly popular worldwide
- Increases natural immunity

1.1.2 History of Herbal Medicine

Archaeological evidence indicates that the use of medicinal plants dates back to the Paleolithic age, approximately 60,000 years ago. Written evidence of herbal remedies dates back over 5,000 years, to the Sumerians, who compiled lists of plants. A number of ancient cultures wrote about plants and their medical uses in books called *herbals*. In ancient Egypt, herbs are mentioned in Egyptian medical papyri, depicted in tomb illustrations, or on rare occasions found in medical jars containing trace amounts of herbs. Among the oldest, lengthiest, and most important medical papyri of ancient Egypt, the Ebers Papyrus dates from about 1550 BC, and covers more than 700 drugs, mainly of plant origin.The earliest known Greek herbals come from Theophrastus of Eresos who in the 4th Century B.C. wrote in Greek *Historia Plantarum,* from Diocles of Carystus who wrote during the 3rd century B.C, and from Krateuas who wrote in the 1st century B.C. Only a few fragments of these works have survived intact, but from what remains scholars have noted a large amount of overlap with the Egyptian herbals. Seeds likely used for herbalism have been found in archaeological sites of Bronze Age

China dating from the Shang Dynasty (1600 BC–1046 BC). Over a hundred of the 224 drugs mentioned in the *Huangdi Neijing*, an early Chinese medical text, are herbs. Herbs also commonly featured in the medicine of ancient India, where the principal treatment for diseases was diet. *De Materia Medica*, originally written in Greek by Pedanius Dioscorides (40–90 AD) of Anazarbus, Cilicia, a Greek physician, pharmacologist and botanist, is a particularly important example of herbal writing; it dominated for some 1500 years.

1.1.3 Scope of Herbal Medicine

The majority of the world's population has access only to traditional, mostly herbal, medicine, so it could be argued that any ailment or disease could be treated with herbal medicine. Some people living in rich industrialised societies, however, have the luxury of being able to choose herbal treatment from a range of healthcare options which include orthodox modern medicine. In this scenario the different healthcare modes are complementary to each other, and the selection of one over another is often a matter of personal choice. Without doubt, orthodox medicine is superior for the treatment of many acute and life-threatening conditions. Herbal medicine, however, has much to offer in the treatment and management of a wide range of conditions that do not constitute medical emergencies.

Many common acute illnesses, such as the common cold, influenza, sinusitis, digestive upsets, insomnia, urinary tract infections and menstrual pain, to mention but a few, can be treated successfully by herbal therapy. It is often in the management of chronic conditions, however, that herbal medicine comes into its own. Herbal medicines are generally well tolerated and associated with only minor side effects. This makes them suitable for use in chronic conditions, where long-term treatment is required. In terms of chronic disease it should be emphasised that herbal medicine is unlikely to cure where conventional medicine has failed, but herbal medicine is often highly successful in managing the symptoms of chronic conditions, resulting in improved quality of life for the patient. The holistic approach to health and disease and the focus on individualised treatment makes herbal medicine an excellent option for the management of chronic disease such as elevated cholesterol levels, problems associated with poor circulation, osteoarthritis, non-insulin-dependent (type 2) diabetes, chronic hepatitis, asthma, anxiety, mild to moderate depression, pre-menstrual syndrome, benign prostatic hyperplasia and chronic skin conditions.

The role of herbal medicine in serious or life-threatening conditions is supportive and complementary to conventional medicine. Herbal medicine cannot substitute for insulin in diabetes, anti-viral drugs in HIV/AIDS or chemotherapy in cancer. What herbal medicine can offer in these conditions is improved quality of life through amelioration of secondary symptoms and side effects of pharmaceutical drugs. Needless to say, only experienced practitioners should treat patients with these conditions, and wherever possible it should be done in concert with the patient's medical practitioner.

Herbal medicine complements a wide range of other therapeutic interventions, in particular nutritional therapy. In the hands of a skilled practitioner herbal medicine can also be successfully combined with conventional medical treatment, with due attention to the potential for drug interactions between herbal medicines and pharmaceutical drugs. As with any case of patient co-management, a superior treatment outcome can be achieved if all practitioners involved are willing to communicate and cooperate.

1.1.4 Upsurge in the use of herbal medicine

In recent times herbal medicine is occupying a strategic position in the general healthcare of people worldwide. Some of its possible reasons are discussed here:

Personal preferences

The use of herbal medicine is an age long tradition which is enshrined in the culture of many societies. In past times, people have relied on it as their primary source of healthcare. Despite modernization and the proliferation of conventional healthcare, people around the globe, especially in Africa and Asia still lay personal preferences on herbal medicine as their first line of therapy, irrespective of their economic, educational and social status. Generally, most of its users have the opinion that herbal medicines are safe and carry no risk or side effects.

Easy accessibility

In rural and tribal communities where access to conventional medical services is either expensive or difficult, the only reliable, easy and quick access to healthcare is the traditional medical practitioners, based on herbal medicine. Even in societies where pharmaceutical products are highly regulated, herbal medicine is easily obtained without medical prescriptions.

Low cost

In many rural communities, the cost of herbal medicines is usually low when compared to the conventional medicines. The traditional medical practitioners, who provide the services, are usually community members as against the long distances to be traversed to reach the conventional medical centers. Most Herbal medicines are prepared or the herbal materials are given to the patients with an oral direction on how to use them effectively. The modalities of payment are usually more flexible as the practitioners, may accept part payments or payment in kind with items such as clothing, chickens, goats, and so on.

Safety and Efficacy

In recent times, there have been increased research activities to verify claims and determine safety and quality control standards for herbal materials and products. The safety and efficacy of some herbal products have been scientifically evaluated to validate claims. The scientific proof of safety and efficacy has contributed to the increasing confidence and popularity of many herbal products.

1.2 Sidhha System of Medicine (SSM)

Siddha medicine is one of the two ancient traditional systems of India. First one is "Ayurveda", which flourished in north India and became popular all over the country and also in abroad, and the second one is "Siddha", which originated from Tamil Nadu, a south-east state of India and practiced mostly in and around areas of its origin.

Siddha system of medicine has been in existence and is being practiced for past 2000 years; however the printed Siddha literatures are available from 18th century onwards. Before that they were documented on palm leaves by different authors.

The word "Siddha" denotes "Siddhi" which means achievement in life arts such as philosophy, yoga, wisdom, alchemy, medicine and above all the art of longevity. The persons, who obtained this Siddhi, were respectfully called "Siddhars". Siddhars can also be called spiritual scientists of Tamil Nadu who explored and explained the reality of nature and its relationship to man by their yogic awareness and experimental findings. They postulated the concept of spiritualism for self-improvement, and the practices initiated by them came to be known as the "Siddha System".

It is believed that Siddhi was obtained by 18 highly experienced and intuitive Siddhars who were believed to cure innumerable diseases. Agasthiyar, one among the 18 Siddhars is believed to have contributed more in the development of Siddha medicinal system, whose life period is approximately between fifth to sixth centuries. He is considered as the "Hippocrates of Siddha medicine and also one of the greatest philosophers of India. Some of his works are still in standard books of medicine and surgery, which are in the daily use among the Siddha medical practitioners.

The SSM is mainly concerned with the development of drugs, which have high potency and long life for their use in future.

1.2.1 Principles of Siddha System of Medicine (SSM)

Siddha system is based on 96 principles *(thathuvams)* which are broadly classified under the following categories:

1. 5 elements *(pancha bhutam)*
2. 5 sense organs *(pori)*
3. 5 functions of sense organs *(pulan)*
4. 5 motor organs *(kanmenthiriyam)*
5. 5 perception of senses with the help of five sense organs *(gnanenthriyam)*
6. 4 intellectual faculties *(karanam)*
7. 1 wisdom of self-realization *(arivu)*
8. 10 channels of life force responsible for the dynamics of prana *(naadi)*
9. 10 vital nerve forces which is responsible for all kinds of movements *(vayu)*
10. 5 visceral cavities *(asayam)*
11. 5 states of the human body or sheath *(kosam)*
12. 6 stations of soul *(aatharam)*
13. 3 regions *(mandalam)*
14. 3 impurities of the soul *(malam)*
15. 3 humours *(tridosham/ tridosha siddhantam)*
16. 3 physical bindings *(eadanai)*

17. 3 cosmic qualities *(gunam)*
18. 2 acts *(vinai)*
19. 8 passions *(ragam)*
20. 5 states of consciousness *(avasthai)*

Many of these principles are found in Ayurveda also but some of them are very unique to Siddha system alone. The common specific features of these two systems are: *tridosha siddhantam*; wind *(vatham)*, bile *(pittam)*, phlegm *(siletuman)*, *pancha bhuta theory*; space (*aakasam*), air *(vayu)* ,fire *(thee)* , water *(neer)* ,earth *(munn)*, *three gunas*; subtle *(sattva)*, activity *(rajas)*, gross *(tamas)*, and six kinds of tastes *(arusuvai)*.

1.2.2 Role of Medicinal plants in Siddha System of Medicines

Numerous plants seem to be used in this system, but there is no adequate data because of multiple reasons. The top two of them are as follows:

1. It is only in ancestral use and inherited by disciple to disciple and generation to generation.
2. It was documented in scholarly Tamil language with numerous cryptic references which differs from the today's common Tamil language and is difficult to understand even by native Tamil people also.

After a long time, some of them were translated into common Tamil language and thereafter, few of these texts were translated into English. Several organizations such as department of Ayurveda, Yoga & Naturopathy, Unani, Siddha and Homeopathy (AYUSH), Government of India; Central Council for Research in Ayurveda and Siddha, New Delhi; and Gandeepam, a non-government organization (NGO) have been established. These organizations have an important role in maintaining and reviving the ancient Indian systems of medicine. In addition, they encourage the scientific studies on these systems.

Alongside, some ethno medicinal studies were also done by some groups of workers in south India [51-53] by gathering the valuable information from the practitioners of rural area near forest region where the people depend mostly on the herbs and have a powerful knowledge of herbal medicines.

Based on these literatures, numerous plants have been observed to play an important role for treating various human ailments in SSM. Some are discussed below:

1. ***Piper longum* L. (Family: Piperaceae); Siddha Name: Thippili**
    a) In SSM, *Piper longum* fruits are usually dried for using in cough, cold and asthma.
    b) It is also used to treat hoarseness, hiccup, colic and flatulence.
    c) It acts as counter irritant, analgesic, haematinic and general tonic.
2. ***Piper nigrum* L. (Family: Piperaceae); Siddha Name: Milagu**
    a) In SSM, the dry unripe fruit powder of *Piper nigrum* along with the honey is given to treat asthma and bronchitis.

b) Its decoction is also used in fever and cough.

c) The milk extract of the fruit is effective in passive cutaneous anaphylaxis.

d) The plant also contains antispasmodic, antioxidant and antibacterial properties.

3. ***Solanum nigrum*** **L. (Family: *Solanaceae*); Siddha Name: Manathakkali**

a) The leaves and fruits of *Solanum nigrum* are used in asthma and bronchitis.

b) The fresh extracts of leaves are also used for inhibiting inflammatory swellings, enlargement of liver and spleen, and in cirrhosis of liver tissue.

c) A glycoprotein from this plant was found to inhibit the DNA binding activities of NF-kB and AP-1 which are one of the important factors for the synthesis of pro inflammatory mediators which cause respiratory diseases like asthma and COPD.

4. ***Euphorbia hirta*** **L. (Family: Euphorbiaceae); Siddha Name: Amman Pachharisi**

Euphorbia hirta is also popularly called as "asthma weed".

a) In SSM, the whole plant is used in treating asthma and bronchitis. Its Anti asthmatic activity may be due to the presence of choline and shikimic acid.

b) The milky latex is applied topically to treat wounds and lip cracks.

c) Recently, its ethanolic extract has been shown to have anti-allergic and anti-anaphylactic effects.

d) It also possesses analgesic, antipyretic and anti-inflammatory properties.

5. ***Crocus sativus*** **L. (Family: Iridaceae); Siddha Name: Kungumapu**

a) The dried stigma and tops of styles are used in treating asthma and cough.

b) Also it is used in multiple ailments such as seminal debility, leucorrhoea, diarrhea and rheumatism.

c) The herb has sedative, antispasmodic expectorant (in dry cough, whooping cough, and bronchitis), stomachic, diaphoretic and antitussive effect.

6. ***Calotropis gigantea*** **L. (Family: Asclepiadaceae); Siddha Name: Erukku**

a) The flowers of *Calotropis gigantea* are used in cough, asthma and loss of appetite.

b) The powdered root bark is soaked in its own milky juice from which bougies are prepared and their fumes are inhaled to treat cough in SSM.

c) The flowering tops pounded and boiled with molasses are given every morning for treating asthma.

d) Its leaf juice is used in external swellings.

7. ***Ocimum sanctum* L. (Family: Lamiaceae); Siddha Name: Thulasi**

a) In SSM, the leaves *of Ocimum sanctum* are used in bronchitis, asthma and chronic cough.

b) It also has wide-ranging therapeutic applications such as in cardiopathy, homeopathy, leucoderma, catarrhal fever, otalgia, hepatopathy, vomiting, lumbago, hiccups, ophthalmia, gastropathy, genito-urinary disorders, ringworm, verminosis and skin diseases.

c) It has been shown to have detoxifying activity.

1.3 Cultivation of medicinal plants

The methods of propagation used in the cultivation of other useful plants apply also to medicinal plants. These include propagation from seeds, cuttings, grafts, buds and other propagules.

Seeds

Most annuals and biennials are propagated from seeds. Considerable care is required while buying the seeds as some of them are immature and in many cases some are sterile. The seed viability (capability of a seed to retain the potential of germination over a certain period of time) generally varies with the plant species. Seeds of *Terminalia bellirica* are viable for a year while the seeds of *Andrographis paniculata* remain viable up to 6 months. Seeds are sown where plants are to be grown, but in most cases it is better to germinate them under glass or in seed boxes and then transplant the young plants when the conditions are most favourable. With many plants by selecting the best seeds or those produced by plants having some special desirable quality, the percentage of phytocompounds has been improved and maintained. It is reported that in the Cinchona plantations in java methods have largely displaced the system of mossing for increasing the alkaloidal percentage.

Seeds of certain medicinal plant species are known to exhibit dormancy and propagation of such species using seeds is not a method of choice. Seeds of *Caesalpinia bonduc, Plantago ovata* and tubers of *Gloriosa superba* exhibit dormancy; however specific treatments are needed which could overcome the dormancy of such seeds.

Cuttings

Cuttings are extensively used for the cultivation of medicinal aromatic and ornamental plants. They are derived either from the over-ground shoots or from the root stocks and rhizomes. In propagating plants from rhizomes, these are cut into pieces having one or two buds, and these pieces are planted separately. Among the medicinal plants which are grown from rhizome cuttings are licorice, ginger etc. Some of the medicinal plants are propagated by hard wood stem cuttings (cutting taken from the branches of the current year's growth of a plant) are *Hibiscus rosa sinensis, Lawsonia inermis* etc. Cuttings produced from tender shoots and branches of the current year's growth that are not too hard, measuring about 20 centimetres with a pair of leaves retained at the tip are commonly known as semi hardwood cuttings. Medicinal plants that are usually propagagted by semi hardwood cuttings are *Adhatoda vasica, Tinospora cordifolia, Eclipta alba, Gymnema sylvestre,* and *Piper* species.

Layering

In this method a vegetative branch is made to root while it is still attached to the mother plant. The rooted branch is later excised and planted as a new seedling. Different layering techniques such as air layering, mound layering, and serpentine layering are commonly practiced. Air layering is frequently followed in Goose berry (Amla), Tamarind and *Ficus* species. Serpentine layering is practiced in those plant species which have drooping long and supple branches like *Tinospora cordifolia, Celastrus paniculatus* etc.

Grafting

Grafting is a propagation technique employed to improve the quality of the nursery stock. It is carried out by bringing together two vegetative parts from two different plants of a same species and joining them together to grow as a single plant. Different techniques of grafting are used for propagating different plants like approach grafting, wedge grafting, epicotyls grafting, softwood grafting. In medicinal plants grafting is seen only with selected species of the genera *Artocarpus, Syzygium, Myristica, Garcinia* etc.

Modified organs

Several plant species including medicinal plants ensure their propagation with the help of modified organs that serve as the propagules.

Medicinal plants such as *Aloe vera, Aegle mormelos, Holorrahena antidysenterica* etc. propagate by suckers (small plantlet that appears around a mother plant from its base and grow as independent plant when separated).

Some medicinal plants propagate by tubers, which are swollen and fleshy underground stem or root. Tuber is the propagating material of many medicinal plants including *Asparagus racemosus, Dioscoria* sp., *Adenia hondala, Gloriosa superba, Ipomoea* sp.

Rhizome, which is the swollen and compressed underground stem with short scaly leaves and vegetative buds, is used for the cultivation of certain medicinal plants such as Ginger, Turmeric, costus etc.

1.4 Harvesting of Medicinal Plants

Medicinal plants should be harvested during the optimal season or time period to ensure the production of medicinal plant materials and finished herbal products of the best possible quality. The time of harvest depends on the plant part to be used.

Guidelines for Harvesting as per World Health Organization (WHO, 2003)

Medicinal plants should be harvested by following sustainable harvesting practices.

1. Medicinal plants should be harvested when they are at the best possible quality for the proposed use.
2. Damaged plants or plants parts need to be excluded.
3. Medicinal plants should be harvested under the best possible conditions avoiding wet soil, dew, rain or exceptionally high air humidity. If harvesting occurs in wet conditions possible adverse effects on the medicinal plant due to increased moisture levels should be counteracted.
4. Cutting devices or harvesters must be adjusted such that contamination from soil particles is reduced to a minimum.
5. The harvested medicinal plants should not come into direct contact with the soil. It must be promptly collected and transported in dry, clean conditions.
6. During harvesting, care should be taken to ensure that no toxic weeds mix with harvested medicinal plants.
7. All containers used during harvesting must be clean and free of contamination from previous harvests. When containers are not in use, they must be kept in dry conditions free of pests and inaccessible to mice/rodents, livestock and domestic animals.
8. Mechanical damage and compacting of the harvested medicinal plant that would result in undesirable quality changes must be avoided. In this respect, attention must be paid to
   a) Over filling of the sacks.
   b) Stacking up of sacks.
9. Freshly harvested medicinal plants must be delivered as quickly as possible to the processing facility in order to prevent thermal degradation.
10. The harvested crop must be protected from pests, mice/rodents, livestock and domestic animals. Any pest control measures taken should be documented.

1.5 Collection of Medicinal plants/materials

Medicinal plants are collected from their natural habitats for their medicinal use and domestic trade. However, certain medicinal plants are over-exploited, due to their high medicinal value. Therefore medicinal plants should be gathered with utmost care from the wild. Sustainable collection will ensure continuous supply of raw materials for pharmaceutical industry, conserve the medicinal plants for the future generation and also protect the natural habitats of such medicinal plants.

Time of collection

Medicinal plants or plant materials should be collected during the appropriate season or time period to ensure the best possible quality as well as quantity **(Table 1).** It is well known that the quantity of biologically active principles varies with the developmental stage of the plant. The best time for collection should be determined according to the quality and quantity of biologically active constituents instead of the total vegetative yield of the targeted medicinal plant parts. The proportion of alkaloid in the leaves of *Hyocyamus niger* and of *Atropa belladona* is largest at the beginning of flowering. Likewise *Datura stromonium* leaves gathered in the morning have higher proportion of alkaloids than those collected in the evening.

Table 1: Calender for the collection of medicinal plant parts

Plant parts	Time of collection
Bulbs	Late autumn, long after flowering and fruit set
Bark	Autumn (after leaf fall) or spring (before leaf renewal)
Roots and Rhizomes	Annuals: Shortly before flowering Biennials: During autumn, following the first year growth Perennials: During autumn, following second or third year growth
Leaves	In dry weather Preferably collection of stems bearing leaves and then their separation Collection in the morning in some cases like Solanaceae family members
Flowers	In dry weather and towards the middle of the day, after dew has dissipated
Seeds and Fruits	Collection when fully grown, ripe or nearly ripe

The collected raw medicinal plant materials should not come in direct contact with the soil. Adhering soil particles from underground plant parts should be removed as soon as they are collected. Collected plant material should be placed in clean baskets, mesh bags or other well aerated containers that are free from foreign matter. The collected plant material should be protected from insects, rodents, birds, animals and other pests. If the collection site is located far away from the processing unit, then the plant material should be air or sun dried prior to transport.

If more than one medicinal plant part is to be collected, they should be gathered separately and transported in separate containers. Cross contamination should be avoided. Collecting implements should be kept clean and maintained in proper condition. Those parts of tools that come into direct contact with the collected medicinal plants should be free from all contaminants.

1.6 Processing of Medicinal plants

Prior to processing, the plant material should be protected from moisture, rain and other deteriorating conditions. Collected plant material should be carefully unpacked at the processing facility. Medicinal plant material should be exposed to direct sunlight only when needed. Plant material that is used in fresh state should be collected, unpacked and delivered for processing quickly in order to prevent microbial and thermal degradation.

After harvesting/collection, the plant material should be immediately stored under refrigeration, in air tight jars or by using other suitable conservation measures. The use of preservatives should be avoided normally and when used they should conform to regulatory requirements of the nation or that particular region.

Medicinal plant materials should be inspected during the processing stages of the production. While inspecting the material, any substandard or foreign material if found should be eliminated. Dried medicinal plant material should be sieved or winnowed to remove damaged materials, soil particles, pebbles or other undesired material. Mechanical devices used should be properly maintained and cleaned.

The primarily processed medicinal plant material should be protected from contamination, decomposition and from insects, rodents and other pests.

The major steps involved in the processing of plants for biological screening are summarised below:

i) Chopping

Collected samples are immediately chopped into small fragments for proper drying and reduce bulk. Thick woody stem, root, twig and other woody parts are chopped into 3-6 cm long pieces and split longitudinally into several sections to facilitate drying. Herbaceous samples and large sized leaves are cut into small fragments. Bark and large fruits are also reduced to pieces. While chopping the plant material, care should be taken to protect the eyes from plant pieces, latex and other injurious conditions, by safety goggles or plastic sheet wrapped around the eyes or face.

ii) Drying

After chopping the plant samples are processed for drying. Under ideal conditions (no threat of rain and dry climate) the samples can be spread under shade without covering.

The medicinal plant material should be dried properly and the moisture content should be kept as low as possible to reduce damage from microbial infestation. Medicinal plant material can be dried by different methods; however, the method and the temperature used for the drying have considerable effect on the quality of the resulting drug. Shade drying is preferred to minimize the loss of colour of leaves and flowers while lower temperature should be employed while drying the plant material containing the volatile substances.

Plant material should be spread out in thin layers on drying frames and stirred or turned frequently. In order to provide adequate air circulation, the drying frames should be made to achieve uniform drying. Plant material rich in essential oil like leaf, flowers etc. should be prevented from bruising. The duration of drying, temperature, humidity and other drying conditions should be determined based on the plant part concerned. Vaccum drying is preferred in certain cases like *Digitalis,* in which pump is used to exhaust the air. Drying herbal plant material directly on bare ground should be avoided. If the surface is concrete, the plant material should be laid on a tarpaulin or other suitable sheeting or cloth.

During rain samples must be kept under shade, properly covered with thick polythene sheet to avoid absorption of moisture. Twigs with leaves and roots take more time to dry than the separated leaves and twigs. In such cases samples should be turned up and down at least twice a day until fully dry. Good examples of such material include the members of Liliaceae (bulbs), Zingiberaceae (rhizome), Crassulaceae, Amarylladiaceae, Orchidaceae, Cactaceae, Euphorbiaceae etc.

iii) Garbling

Before packing the material, the extraneous matter such as unwanted parts of the same plant or grasses or herbs or any other foreign matter should be removed from the leafy, tender stems, rhizomes, tubers, roots or aerial portion samples.

iv) Packing and transportation

Samples are susceptible to deteriorate from the absorbed moisture, therefore they are packed in water-proof round canvas or jute kit bags. One should be careful that the samples should be dried upto a certain stage so that there is least chance of spoiling during transportation. In general, the plant sample must reduce to half the weight of the material originally collected.

For half dried samples, use of jute kit bags is advantageous in that they get some air to circulate thus avoiding any rotting. All samples on reaching the laboratory are then again kept open under shade for few days and then repacked in order to avoid contamination.

v) Grinding

The grinding operation should be done by experienced hands who are properly trained. Proper care should be taken while grinding the plant material to avoid hazard damage to eyes, injury and inhalation of fine powder or other allergic reactions.

vi) Storage

Proper preservation is necessary for maintaining the high quality of plant sample. Care should be taken to protect samples from the rodents and insects. Storage room should be kept cool and dark with no risk of moisture being absorbed by the material. Moisture not only causes increase in the weight of material but reduces the percentage of active principles. Moisture also favours microbial growth and thus leads to destruction of the samples. Destruction of samples by insects is generally minimized effectively by using few drops of carbon tetrachloride or chloroform. To get ideal results, the storage period should be the minimum possible. Following measures should be taken while storing the medicinal plant material for drug development:

1. Storage facilities for medicinal material should be well aerated, dry and protected from light, and, when necessary, be supplied with air-conditioning and humidity control equipment as well as facilities to protect against rodents, insects and livestock.

2. The floor should be tidy, without cracks and easy to clean.
3. Medicinal material should be stored on shelves which keep the material a sufficient distance from the walls.
4. Measures should be taken to prevent the occurrence of pest infestation, mould formation, rotting or loss of oil; and inspections should be carried out at regular intervals.
5. Continuous in-process quality control measures should be implemented to eliminate substandard materials, contaminants and foreign matter prior to and during the final stages of packaging.
6. Processed medicinal plant materials should be packaged in clean, dry boxes, sacks, bags or other containers in accordance with standard operating procedures and national and/or regional regulations of the producer and the end-user countries.
7. Materials used for packaging should be non-polluting, clean, dry and in undamaged condition and should conform to the quality requirements for the medicinal plant materials concerned. Fragile medicinal plant materials should be packaged in rigid containers.
8. Dried medicinal plants including essential oils, should be stored in a dry, well-aerated building, in which daily temperature fluctuations are limited and good aeration is ensured
9. Fresh medicinal plant materials should be stored at appropriate low temperatures, ideally at 2-8°C; frozen products should be stored at less than -20°C.
10. Small quantity of crude drugs could be readily stored in air tight, moisture proof and light proof container such as tin, cans, covered metal tins or amber glass containers.
11. Wooden boxes and paper bags should not be used for storage of medicinal plants.

1.7 Marketing of Medicinal Plants

Medicinal and aromatic plants are traded both as commodities and processed formulated drugs. The wide range of products obtained from medicinal plants includes herbal medicinal products (conventional and traditional); herbal drinks; food/dietary supplements; foodstuffs; cosmetics, perfumes etc. Formulated drugs are prepared as per the descriptions in the ayurvedic, unani and siddha systems of medicine. Currently, the market for herbal products throughout the globe is approximately around US$60 billion per year with an annual growth rate of 7%. WHO estimated that 80% population of developing countries relies on traditional medicines, particularly based on plant based drugs, for their primary healthcare needs.

In china, India and Nepal, medicinal plants are an important revenue generating resource especially for economically marginalized and indigenous people. Increasing interest by multinational pharmaceutical companies and domestic manufacturers of herbal medicines is contributing significant economic growth of the global medicinal plants sector. In 1997, 3500 tonns of bark of *Prunus africana* (valued US$220 million) was exported from Africa. The global market potential of Aloe vera for treating burns and as a constituent of skin creams and cosmetics is estimated in the billions of dollars. Global aloe-vera market is anticipated to expand from USD 1,528.5 million in 2015 to USD 2,344.2 million in 2021, exhibiting a compound growth rate of more than seven percent. From the three Himalayan countries including Nepal, Bangladesh and Bhutan, several thousand tonns of medicinal and aromatic plants are extracted from forests, providing earnings that run into millions of dollars each year.

In Nepal, it is estimated that every year 20,000 tonns of medicinal plants worth US$18-20 millions are traded and approximately 90% of this collection is exported mainly to India in raw form. Medicinal plant trade in India is substantial with total turnover of Rs. 2300 crores of ayurvedic and herbal products. The annual revenue stream from the three major Indian systems of medicine i.e. Ayurveda, Siddha and Unani, is estimated at more than half a billion dollars.

With approximately 6600 medicinal plants, India is the second largest exporter of medicinal plants in the world, following China. They produce over 70 % of herbal medicine's demand across the globe. India alone exports the medicinal plants at the cost of 10,000 crore rupees per year. India exports about 75-80% of the crude drugs to six developed countries including USA, Germany, France, UK, Japan and Switzerland. The principal herbal drugs that have been finding a good market in foreign countries include Aconite, Aloe, Belladona, Cinchona, Cassia tora, Dioscorea, Digitalis, Ephedra, Ergot, Hyoscymus, Liquorice, Opium, Papain, Podophyllum, Rauwolfia, Senna, Stramonium etc.

At present, the demand for medicinal plants is increasing throughout the globe and in developing countries particularly because of the popularization of the traditional medicines. Some medicinal plants are highly exploited and are practised in large quantities, such medicinal plants are highly demanded in the market and are known by prioritized medicinal plants. The annual demand and the growth rate of such medicinal plants in India are presented in **Table 2.**

Table 2: Prioritized medicinal plants in India

Plant Species	Demand (tonnes)	Annual Growth Rate (%)
Indian aconite	3427	30.0
Amla	41783	22.5
Bacopa	6622	20.1
Guggul	1505	19.2
Sandalwood	1073	19.1
Atis	448	18.4
Holy basil	5403	17.9
Chlorophytum	61	17.2

Plant Species	Demand (tonnes)	Annual Growth Rate (%)
Long pepper	6280	16.3
Liquorice root	1360	15.9
Indian barberry	1829	15.5
Flame lily	101	15.4
Shatavari	16659	15.1
Asoka tree	10724	15.0
Picrorhiza	317	12.9
Sarpagandha	589	11.6
Bhui-amla	2985	10.5
Chirata	1285	10.0
Bael tree	7085	9.6
Ashwagandha	9128	9.1
Giloy	2933	9.1
Costus (Kuth)	1826	8.9

1.8 Conservation of medicinal plants

1.8.1 Threats to medicinal plants and the need for their conservation

Factors like degradation of habitat, forest decline, destructive collection of medicinal plant species, invasion of exotic species, high susceptibility to diseases, industrialization, over exploitation, excessive use of agrochemicals, natural and manmade calamities, genetic erosion etc. cause threat to many medicinal plants. To meet the requirements of expanding markets for healthcare products and needs of growing populations, large quantities of medicinal plants are harvested. In India large numbers of medicinal plants are extracted to meet the increasing demand for domestic consumption and for the export. As a result, the medicinal plant wealth is depleting rapidly. In South India alone, it is estimated that approximately 70-80 out of 300 medicinal plants are either endangered or threatened. Hence there is a need for conservation, cultivation and the maintenance of medicinal plants for the future use.

1.8.2 Conservation strategies for medicinal plants

It is essential that feasible strategies should be employed to conserve the populations and genetic resources of medicinally important species to avoid further loss. The two main strategies for the conservation of medicinal plants are *ex situ* (protection of species outside their natural habitats) and *in situ* (in their natural surroundings) **(Fig.1)**.

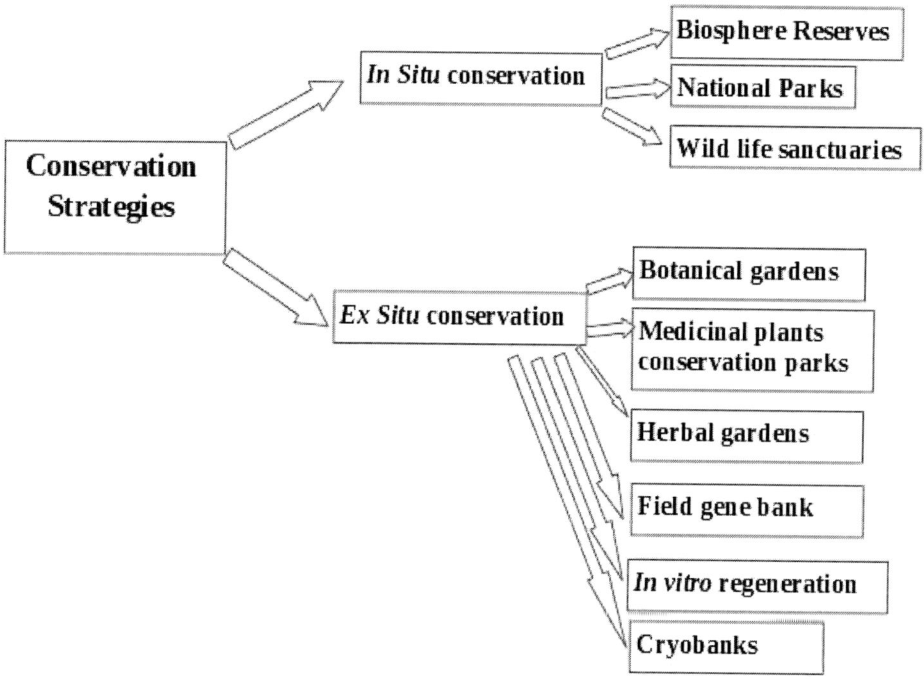

Fig. 1: Different strategies for the conservation of medicinal plants

1.8.3 *In situ* conservation

In situ conservation involves maintaining genetic resources in their natural habitats i.e., within the ecosystem to which it is adapted. The major operational steps for establishing *in situ* gene banks for the conservation of medicinal plants include: threat assessment, establishment of a network of medicinal plant forest reserves, involving local stakeholders, botanical, ecological, trade and ethno-medical surveys, assessing intraspecific variability of species, designing species recovery programmes, establishment of a medicinal plant seed centers etc. *in situ* conservation projects can be successful only by cooperation and involvement of local people.

Medicinal plant conservation areas (MPCAs) and medicinal plant development areas (MPDAs) are the people-oriented approaches for the in situ conservation of medicinal plants. MPCAs are the strictly protected areas where people are not allowed to exploit the medicinal wealth while in MPDAs people are allowed to harvest the medicinal plants sustainably.

1.8.4 *Ex situ* conservation

Ex situ conservation, involves conservation of biodiversity outside the native or natural habitat where the genetic variation is maintained away from its original location. The *ex situ* genetic conservation fulfills the requirement of present or future economic, social and environmental needs. Conservation also includes propagation and assessment of molecular diversity

Conservation of medicinal plants include a combination of methods, depending on factors such as geographic sites, biological characteristics of plants, available infrastructure and network having an access to different geographical areas, human resources and number of accessions in a given collection.

The main strategies of *ex situ* conservation of medicinal plants are:-

1. **Botanical gardens**

   Botanical gardens are the small areas where various plant species collected from different regions are grown in blocks. They play an important role in the conservation of threatened and endangered plant species. Botanical gardens are generally attached to the botany department in colleges and universities while others are administered by research institutions. Several botanic garden activities are important for the conservation of medicinal plants around the world, including:

   Horticulture and cultivation skills allow botanic gardens to grow plants that might be lost in the wild.

   Education is the basic strength of botanic gardens that allows them to communicate the importance of medicinal plants and their conservation.

   Linking plants with the health of people encourage the sustainable use of medicinal plant sources.

2. **Medicinal plants conservation parks (MPCPs)**

   MPCPs are the areas owned by non-government organizations, where medicinal plants are grown for the public interests. In these parks plants having potential traditional importance are frequently grown by the local people under the guidance of NGOs. MPCPs are the essential source to confer the medicinal uses of various plants to the local people. In India 108 medicinal plants conservation areas were established since 1993. These areas are not profit oriented but provide raw material for use as traditional medicine for local people.

3. **Herbal gardens**

   Herbal gardens are established to popularize the usefulness of commonly available and frequently used medicinal plants among the various stakeholders. On the other hand it is a way of conserving rare, endangered and threatened species of medicinal plants. These include herbal garden of national and state importance as well as at the level of research institutions, government aided schools, universities, colleges and homes.

   In home herbal gardens medicinal plants are grown for primary healthcare at the household level. About 148,000 home herbal gardens were established in south India during 2007-2008. Nearly 20 different species of medicinal plants are grown in home herbal gardens. Herbs are usually grown in separate blocks while shrubs and trees are grown towards the periphery. Surplus medicinal plants are collected sustainably and sold to the local traders. The old aged traditional knowledge system is a way of life in Indian households and herbal home garden render valuable healthcare system.

4. **Field gene bank**

This is a method of growing plants for the conservation of genes. For this purpose, an ecosystem is created artificially, needs more land, adequate soil, weather etc. The first field gene bank was established at Kerala in a forest region of the Western Ghats Mountains by Tropical Botanical Gardens and Research Institute (TBGRI). This field gene bank has 250 medicinal and aromatic plant species with nearly 100 endemic, rare and endangered plant species.

In field gene banks the plant genetic resources are kept as live plants by continuous growth and maintenance. They are used when the germplasm is either difficult to conserve as seeds (i.e. no seed formation, recalcitrant seeds or seed production takes several years) or the crop is reproduced vegetatively. Maintaining plants in field genebanks is costly and this method of conservation is usually used when there are no available alternatives or the storage period of other methods is very short. Field gene bank conservation involves the collection of materials from farmer's fields and gardens or even from wild locations and transferring them to second site where they can be planted and monitored. Field genebanks are mostly used for the conservation of clonal crops, vegetatively propagated genotypes that commonly produce variants (genetic variation) as they can be more easily identified. Germplasm of important plant species are conserved through this method. Field gene banks provide easy and ready access to conserve material for research as well as for their use.

5. ***In vitro* regeneration**

In vitro regeneration includes plant/explant growth, maintenance under disease free condition, retention of regenerative potential, genetic stability. It offers a number of advantages over the *in vivo* method:

a) Economical in storage space and time

b) Maintaining of species for which seed preservation is difficult or unsuitable and

c) Disease-free transport and exchange of germplasm.

In vitro multiplication protocols for fast propagation of a number of IUCN Red listed medicinal and aromatic plant species which are rather difficult to propagate through conventional means would be very useful. Usually, shoot tips or axillary buds are cultured on a nutrient medium containing (i) high levels of cytokinins or (ii) low concentrations of auxin coupled with high-cytokinin content. Somatic embryos, or even axillary buds are encapsulated in hydro soluble gels to form 'artificial seeds' for rapid propagation of the species.

Basic culture technique

The *in vitro* propagation starts with the selection of explant from a healthy and vigorous mother plant. Initially the explant is surface sterilized and transferred into a suitable nutrient medium under controlled conditions.

The explant undergoes multiplication by repeated sub-culturing. For rooting stage, the cultural conditions are modified by changing the nutritional and growth regulator composition. The *in vitro* plants are then weaned and hardened and then transferred to an appropriate substrate (sand, peat etc.) under green house.

6. **Cryobanks for conservation**

   Cryopreservation of plant cells and meristems is an important tool for long term storage of germplasm without genetic alteration using a minimum space and maintenance. Cryobanks are basically meant for storage of germplasm in the form of cell suspension, callus, axillary buds, somatic embryos etc. For long term preservation, cryogenic storage at ultra low temperatures in liquid nitrogen (-150 to -196°C) is one of the best method of choice.

   Encapsulation/dehydration is another new technique that offers practical advantages. It is based on the technology originally developed for production of synthetic seeds, i.e., somatic embryos encapsulated in a hydrosoluble gel. Several types of *in-vitro* raised materials such as meristems/shoot tips, cell suspensions, protoplasts, somatic embryos and pollen embryos of medicinal and aromatic species have been studied from the cryopreservation perspective. Meristem cultured calli from medicinal and aromatic plants are inoculated in screw-capped tubes and treated with cryoprotective agents like dimethylsulfoxide, polyethylene glycol, polyethylene oxide etc. Tubes are then kept in a freezer at freezing temperature (-79 ^{0}C). Tubes are then wrapped with thick plastic cover and kept in cryofreezer, filled with liquid nitrogen at -195 ^{0}C. The medicinal plant material can be conserved by this approach for several years without any genetic alteration.

7. **Seed storage modules**

   Seeds, being natural perennating structures of plants, are best suited for storage. By suitably altering their moisture content (5-8%), they can be maintained for relatively long periods at low temperatures (-18 °C or lower). However, in several species, rhizome/bulb or some other vegetative parts may be the site of storage of active ingredients, and often, such species do not set seed. If seeds set, they may be sterile or recalcitrant i.e., intolerant of reduction in moisture or temperature, thus unsuitable for storage. It is now possible to store materials other than seed, such as pollens or clones obtained from elite genotypes/cell lines with special attributes, *in vitro* raised tissues/organs, or, genetically transformed material.

1.9 Micropropagation of Medicinal plants

Micropropagation is the in vitro propagation of plants from cells, tissues and organs such as roots, shoot tips, leaves etc. through tissue culture method. The use of tissue culture for micropropagation was initiated by G. Morel (1960). Plants obtained from tissue culture are called "*microplants*". Microplants can be generated in three different ways:

a) From pre-existing shoot buds or meristems which are encouraged to proliferate.

b) Following shoot morphogenesis when new shoots are induced to form in unorganised tissues.

c) Through the formation of somatic embryos (somatic embryogenesis).

Fig. 2: General micropropagation technique

Micropropagation technique

Micropropagation of medicinal plants involves some important stage which are illustrated in Fig. 2 and discussed below:

1. **Preparation of explant**: Micropropagation is generally initiated from pieces of a plant, termed explants, which may consist of pieces of organs (root, shoot tips, leaves) or specific cell types like pollen or embryo or endosperm etc. Explants are surface sterilized by treatment with a disinfectant solution like 1-2% sodium hypochlorite, 9-10% calcium hypochlorite or 70% ethanol.

2. **Formation of callus**: Callus is an unorganised, undifferentiated mass of cells, formed when explants multiply in a disorganised way. Callus can be initiated in vitro by placing explants onto a growth supporting medium under sterile conditions. During this process explants give rise to new tissue, composed of meristematic and unspecialized cell types.
3. **Shoot development**: An undifferientiated callus can be regenerated into a shoot system by simply supplying growth regulators in a required amount. Shoots are generally initiated and multiply when the cytokinin/ auxin ratio is high.
4. **Root formation**: The multiplied shoots are excised and transferred to another culture medium to induce rooting. Roots are generally induced when the culture medium is supplemented with high auxin/cytokinin ratio.
5. **Transfer to a glasshouse**: In vitro raised plantlets are then transferred to pots containing sand and soil mixture in a glasshouse for acclimatization.

1.9.1 Micropropagation of Ashwagandha *(Withania somnifera)*

Significance

Withania somnifera, popularly known as Ashwagandha grows on both lower as well as higher altitude as a rare plant, thus needed to be conserved. It should be propagated on lower altitudes to have its commercially significant products like secondary metabolites. Secondary metabolites of *Withania somnifera* have got potential application as therapeutics. Ashwagandha forms active ingredient of more than 100 Ayurvedic, 30 Sidha and a good number of Unani medicines. As germination of seeds in this plant is very poor, it is better to grow the plant, *in vitro* conditions from various explants. The technique of micropropagation is applied with the objective of enhancing the rate of multiplication. Through the culture over a million of plants can be grown from a small piece of plant tissue within 12 months.

Micropropagation Methods

Shoot multiplication was achieved *in vitro* from shoot tips of aseptically germinated seedlings of *Withania somnifera* using low concentrations of 6-benzyladenine (BA) **(Fig. 3).** Maximum number of shoots were obtained when 2.3 μM 2, 4-dichlorophenoxyacetic acid (2,4-D) or 2.5 μM indole butyric acid (IBA) was added to medium, containing 4.4 μM BA during initiation of shoot multiplication. Direct multiple shoot initiation was also obtained from germinating seeds in the presence of BA alone. Rooting was successful in excised shoots grown on growth regulator-free MS medium. Rooted shoots were successfully established in soil in a greenhouse. Direct rhizogenesis was obtained from leaves of *Withania somnifera* by using an IBA dip treatment. The segments dipped in IBA formed roots along the midrib region of the abaxial surface when placed on Murashige and Skoog's (MS) basal medium without plant growth regulators. Maximum rooting (85.3 percent of the cultures) occurred when the leaf segments were placed on ½ strength MS medium after a dip treatment with 100 mg/liter IBA solution for 20 min. The average number and length of roots were 32.3 per culture and 5.6 cm, respectively. Only 20% of the cultures produced roots if explants were grown on full-strength MS medium supplemented with IBA.

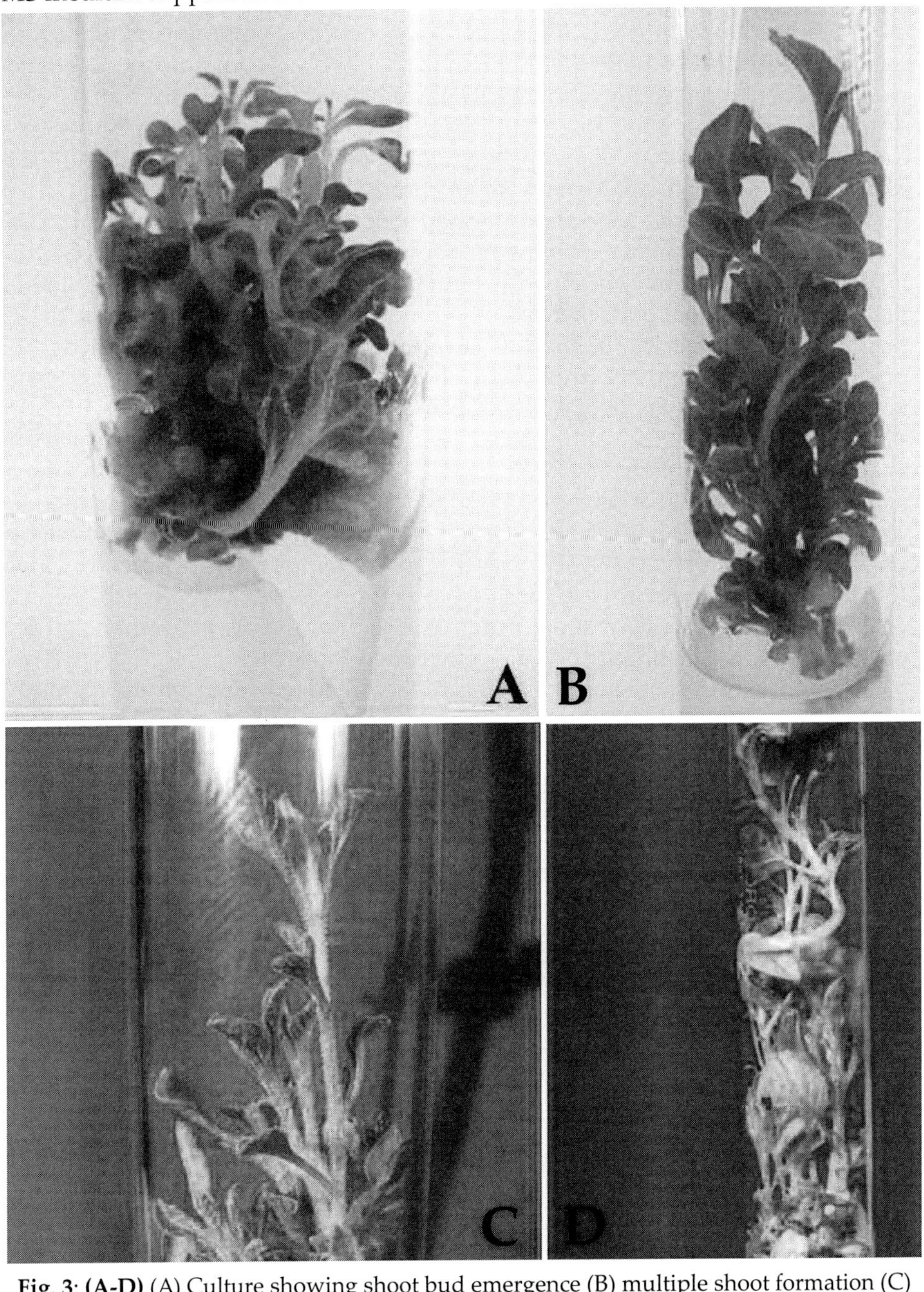

Fig. 3: (A-D) (A) Culture showing shoot bud emergence (B) multiple shoot formation (C) Culture showing direct regeneration of multiple shoots (D) Multiple shoot formation

An efficient protocol was developed for large scale propagation using seed as explant with MS medium supplemented with BAP 0.6 mg/l and IAA 0.4 mg/l. About 90% rooting was achieved with 0.4 mg/l IBA and 0.4 mg/l IAA. Shoots were also regenerated from node, internode, hypocotyl and embryo explants. Direct regeneration of shoot buds were obtained in MS basal medium supplemented with various concentrations of either benzyladenine (BA) or thidiazouron (TDZ) depending on the explant. Nodal explants formed multiple shoots both from pre-existing and de novo buds on Murashige and Skoog's medium (MS) containing 0.1–5.0 mg l^{-1} BA and a ring of de novo shoot buds on MS medium containing 0.2 and 0.3 mg l–1 TDZ. Internodal explants formed shoot buds on MS with 1.0 and 5.0 mg l^{-1} BA while the hypocotyl explants gave rise to multiple shoots only on MS with 0.5 mg l^{-1} BA. Isolated embryos gave rise to many shoot buds on MS with 0.2 and 0.3 mg l^{-1} TDZ. The shoot buds elongated and rooted either on MS medium with 0.01 mg l^{-1} BA or on half strength MS medium lacking growth regulators, which depend upon the growth regulator used in the shoot bud induction medium. Except for the embryo-derived plantlets, all other plantlets could be acclimatized with 100% success. Callus cultures were initiated from nodal segments on Murashige and Skoog medium supplemented with 2,4-D, BAP and Kn. The highest frequency (85%) of organogenic callus induction was observed in MS medium containing 1 mg/ L/ha BAP and 2 mg/L/ha Kn. Development of adventitious shoots occurred when the calli were subcultured in MS medium supplemented in the BAP and Kn. Shoots differentiated best (80%) from node derived callus on MS medium containing 1 mg/L/ ha BAP and 2.5 mg/ L/ha Kn. Regenerated shoots rooted best on MS medium containing IBA and Kn (1 mg/L/ha). Plantlets were transferred to pots containing sand and soil mixture, acclimatized in a culture room and finally rooted plants were transferred to soil. Direct rooting from leaf explants of *Withania somnifera* was achieved on half strength Murashige and Skoog's medium supplemented with 15g/l sucrose and different concentrations of growth regulators. Basal medium supplemented with 2.85 μM indole acetic acid and 9.85 μM indole butyric acid achieved maximum number of roots with 100% response. The roots were cultured on MS liquid medium for the establishment of root-organ culture with the same plant growth regulators.

1.9.2 Micropropagation of Neem (*Azadirachta indica*)

Neem seeds lose viability in a fortnight of storage. Furthermore, the seed and seedling vigour, diminishes well before the seeds start losing viability leading to abnormalities. This has led to problems in plant genetic resource conservation, establishment of large-scale plantations, and species trials. Clonal propagation offers the advantage of potentially rapid multiplication.

The *in vitro* propagation of Neem is carried out by different plant tissue culture approaches. Some of the successful methods are: (i) induction of the formation of multiple shoots from axillary buds (ii) direct regeneration from somatic tissues (iii) direct somatic embryogenesis (iv) indirect somatic embryogenesis etc.

The overall *in vitro* propagation of shoots and nodal segments of Neem is depicted in **Figure 4**.

Fig. 4: ***In vitro*** **culture of** ***A. indica*** (A) Nodal segments showing shoot elongation (B) Shoots with mild senescence (C) Shoots grown in ventilated flasks (D) Early acclimatization in cups covered with plastic bags (E) After acclimatization without plastic bags (F) Acclimatized explant after 60 days in a greenhouse

Some of the important culture techniques successfully applied to *A. indica* are discussed below:

Leaf culture

Plantlet regeneration has been reported in neem from leaf explants of a mature tree. Shoots were successfully produced by culturing leaf explants on Murashige and Skoog medium containing benzylaminopurine (1 mg/l), kinetin (0.8 mg/l) and adenine sulphate (6 mg/l). 80 shoots in 32 weeks could be produced from a single leaf explant of a mature neem tree. 55% of these shoots rooted on MS medium containing Indolebutyric acid (1 mg/l) and all grew on transfer to soil.

Anther culture

The combination of 4 ppm NAA with 2.56 ppm kinetin in MS medium was found to be the best for callus formation from anthers after 30-35 days of incubation. The callus developed into shoot buds on MS medium supplemented with 1 ppm of BAP. Addition of 0.25 mg/l silver nitrate produced longer shoots with broader leave. The excised shoots developed healthy roots upon transfer to half-strength MS medium containing 1 ppm IAA after 15 days of incubation.

Bud culture

Neem was regenerated by culture of buds from crown branches of a mature tree (15-year-old), basal-sprouts of another mature tree (15-year-old) and a single juvenile plant (1-yearold). Cultures derived from these three different sources showed significant variation in *in vitro* response. DKW (Driver and Kuniyuki) medium (with 0.22μM BA) was significantly better than MS (Murashige and Skoog's) medium for shoot proliferation. Shoot cultures of crown branch origin did not elongate and eventually died after the third subculture. In the presence of 4.9μM IBA in half-strength DKW, 90% of the shoots and 100% of basal-sprout and seedling explant origin, formed roots. Plantlets from both explant types showed 90% survival after acclimatization.

1.9.3 Micropropagation of Tulsi (*Ocimum sanctum Linn.*)

Significance

The conventional method for propagation of *Ocimum sanctum* is through seeds. However, poor germination potential and lack of vegetative propagation restricts its multiplication. Micropropagation is an effective tool for rapid multiplication of species in which conventional methods have limitations. There is one major report on this plant in which micropropagation has been achieved using young inflorescence explants. The young inflorescence has been well recognized as a potent source for the micropropagation of many plant species. The advantage of inflorescence explants is the ease of material sterilization and conservation of the mother plant utilized to initiate the culture.

The protocol established for the *In vitro* multiplication of *Ocimum sanctum* shoots is illustrated in Fig. 5 and summarized below:

Explants and sterilization

Young inflorescences (0.6–1.2 cm long) were excised from the plant and thoroughly washed. The inflorescences were surface sterilized in 0.2% (w/v) aqueous mercuric chloride solution and finally the excess sterilant was removed by repeatedly washing in sterile distilled water under aseptic conditions.

Culture conditions

Sterilized explants were cultured individually under aseptic conditions on Murashige and Skoog medium containing 3.0% sucrose and 0.8% agar. This served as a basal medium. MS medium was also supplemented with various growth regulators (IAA, TDZ) either individually or in various combinations. Cultures were maintained at 25°C and 55% relative humidity under continuous light (1000–2000 lux).

Rooting of shoots and acclimatization of plantlets

Regenerated shoots were excised and transferred individually to MS medium without growth regulator supplements. Approximate 92% shoots developed 8–12 roots within 2– 3 weeks. *In vitro* raised plantlets were subsequently transferred individually to plastic cups filled with vermiculite and garden soil (1:1). These plantlets resumed normal growth and developed healthy leaves 2 weeks after transplantation. Out of these 85% plantlets survived. These plantlets were then transferred to the pots containing soil and they were finally transferred to the field where they grew normally. All these micropropagated plants were phenotypically similar to the mother plant.

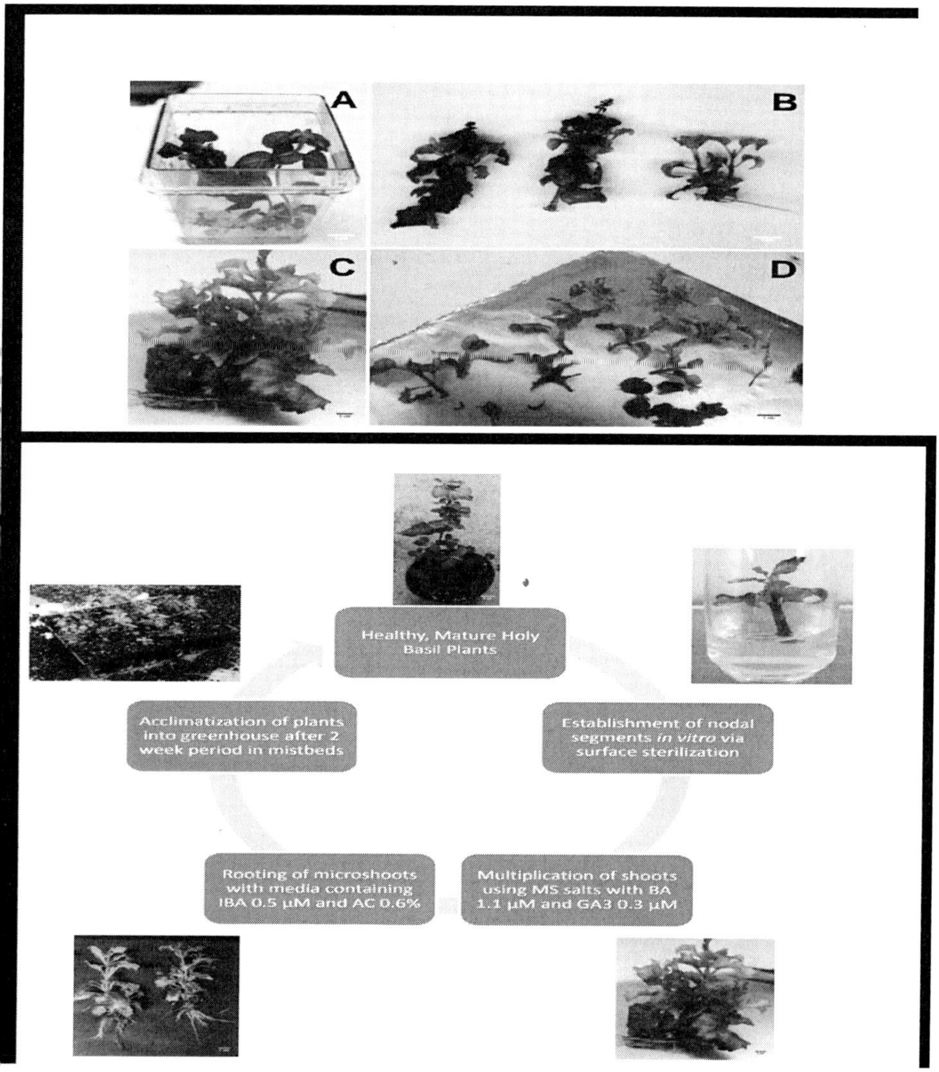

Fig. 5: Photographs demonstrating shoot multiplication (A) *Ocimum sanctum* single shoot formation (B-D) *O. sanctum* shoot multiplication

UNIT - II

Phytochemistry

Phytochemistry is a diverse discipline which deals with a variety of organic substances or natural products accumulated in plants, called "phytochemicals". These chemical compounds are formed during the plants normal metabolic processes and are often referred to as "secondary metabolities".

There are several classes of phytochemicals, including alkaloids, anthraquinones, flavonoids, coumarins, glycosides (cardiac glycosides and cyanogenic glycosides), gums, polysaccharides, phenols, saponins, tannins, terpenes and terpenoids present in the plants. Most of the phytochemical constituents present in medicinal plants are the precursors for the synthesis of useful drugs.

2.1 Phytochemicals and active principles

Phytochemicals which are biologically active and have some thereupatic action are known as active principles. Active principles present in the medicinal plants have been exploited in healthcare industry for the treatment of various ailments over the years, due to their diverse therapeutic potential (Table 3). Some of the active phytochemicals enhance their own survival while others are active in other organisms.

Table 3: Plant based active secondary metabolites and their therapeutic action

S. No.	Active Principle	Therapeutic potential
01	Ajmalicine	Antihypertensive
02	Ajmaline	Anti-arrhythmic
03	Atropine	Muscle relaxant
04	Azadirachtin	Insecticidal
05	Baccharine	Antineoplastic
06	Berberine	Antidiabetic
07	Betulinic acid	Antitumor
08	Bruceantine	Antineoplastic
09	Caffeine	Stimulant
10	Catechin	Antioxidant
11	Chrysophanic acid	Anticancerous
12	Codeine	Analgesic

S. No.	Active Principle	Therapeutic potential
13	Colchicine	Gout suppressant
14	Digitoxin	Cardiovascular
15	Digoxin	Cardiotonic
16	Diosgunin	Antifertility
17	Ellipticine	Antineoplastic
18	Emodin	Laxative
19	Fagaronive	Antineoplastic
20	Gsaline	Antineoplastic
21	Hesperidin	Antioxidant
22	Hypercin	Antidepressant
23	Isorhamnetin	Anti-inflammatory
24	Jandicine N-oxide	Antineoplastic
25	Kaempferol	Antioxidant
26	Maytansive	Antineoplastic
27	Morphine	Analgesic
28	Nicotine	Insecticidal
29	Papaverine	Vasodilator
30	Phytoecdysones	Insecticidal
31	Podophyllotoxin	Antineoplastic
32	Pyrethrins	Insecticidal
33	Quercetin	Antioxidant/Anti-inflammatory
34	Quinine	Antimalarial
35	Rosamarinic acid	Antioxidant
36	Rotenoids	Insecticidal
37	Rutin	Antioxidant
38	Scopolamine	Sedative
39	Suolpolanine	Antihypertensive
40	Taxol	Antineoplastic
41	Terpendids	Spasmolytic
42	Thalicarpine	Antineoplastic
43	Tlarringtovinl	Antineoplastic
44	Tripdiolide	Antineoplastic
45	Vinblastin	Antineoplastic
46	Vincristine	Antineoplastic
47	Vitexin	Antineoplastic
48	3-Doxycolchicine	Antineoplastic

S. No.	Active Principle	Therapeutic potential
49	9-methoxyellipticine	Antineoplastic
50	Galangin	Anti-mutagenic

2.2 Classification of Phytochemicals

Active chemicals present in plants have different nature and chemical structure and belong to various phytocompound classes. The major active principle classes of plants are being discussed.

1. **Alkaloids**

   Alkaloids are a diverse group and comprise the largest single class of secondary metabolites. There is no one specific definition of alkaloid, but alkaloids are those basic compounds which contain one or more nitrogen atoms, usually in combination as part of a cyclic system. The most common precursors of alkaloids are the amino acids and are found in about 20% of plant species. Being often toxic, alkaloids play a major defensive role against herbivores and pathogens in the plants. Due to their strong biological activities, approximately 12,000 known alkaloids have been developed as tonic, stimulants, narcotics and poisons. The first alkaloid identified from the plants was morphine, isolated from *Papaver somniferum* (opium) in 1806. Morphine is probably the most extensively used analgesic for the treatment of acute pain associated with injury, neuropathic disorders and cancer. Besides morphine, plant-derived alkaloids currently in clinical use include codeine (analgesic), vinblastine (anti-neoplastic), colchicines (gout suppressant), tubocurarine and papaverine (muscle relaxants), ajmaline (anti-arrhythmic) and scopolamine (sedative). Other well known alkaloids of plant origin include caffeine (*Coffea* sp.) nicotine (*Nicotiana tabacum*), cocaine (*Erythroxylum coca* var. *coca*), heroin (morphine derivative), berberine (*Berberis vulgaris*), atropine (*Atropa belladonna*), senecionine (*Senecio jacobaea*), cytisine (*Cytisus laburnum*), strychnine (*Strychnos nux-vomica*), aconite (*Aconitum napellus*), quinine (*Cinchona officinalis*) etc.

   Quinine alkaloid is highly used in the treatment of malarial infections of resistant strains. Diterpenoid alkaloids from the members of Ranunculaceae family are commonly found to have antibacterial and antifungal properties. Solamargine (*Solanum khasianum*), a glycoalkaloid is useful against HIV infection and intestinal infections. Berberine is essentially effective against trypanosomes and plasmodia. Sanguinarine alkaloid commonly obtained from Papaveraceae and Fumariaceae members is an antimicrobial and anti-inflammatory agent which has been commonly used to treat gingivitis and plaque formation.

   The "belladonna series" is a group of drugs with similar physiological action obtained from Solanaceeae members viz. *Atropa belladonna, Datura stramonium* and *Hyoscyamus niger*. Belladonna is the most familiar name in the list and today the mydriatic property of atropine is highly valuable in ophthalmology. *Datura stramonium* has long been used as a part of

Ayurvedic preparations due to the presence of alkaloids. *Hyoscyamus niger*, commonly known as "henbane" yields atropine, hyoscymine and scopolamine.

Cocaine isolated from *Erythroxylum coca* has a place in history of relief of pain. The other important alkaloid is quinine, obtained from *Cinchona*. Quinine is a high molecular weight alkaloid used as an effective insecticide. Work on roots of *Rauvolfia serpentina* resulted in the isolation of another important alkaloid namely reserpine. Reserpine is now sold in the market as Serpacil.

2. Phenolics

The term phenolic compound comprises a wide range of plant substances which possess in common an aromatic ring with one or more hydroxyl (OH) substituents. Over 8000 phenolic structures have been reported and they are widespread throughout the plant kingdom. Phenolic compounds range from simple, single aromatic-ringed compounds to large and complex polyphenols. Phenolics can be discussed under following groups.

- ***Flavonoids***

  Flavonoids are phenolic compounds having fifteen carbons, with two aromatic rings connected by a three-carbon bridge. Flavonoids are usually present in all vascular plants but some classes are more widely distributed than others; while flavones and flavonols are universal, isoflavones and biflavonyls are found in only a few plant families. In plants, flavonoids are involved in important diverse processes such as UV protection, pigmentation, stimulation of nitrogen-fixing nodules and disease resistance. Anthocyanins, Proanthocyanidins, flavones, glycoflavones, biflavonyls, chalcones, flavonones, isoflavones etc. are the different flavonoid classes.

  Flavonols are the most widespread flavonoids, found in most fruits, vegetables and beverages. Myricetin, quercetin, isorhamnetin and kaempferol are most commonly found flavonols in plants.

  Flavonoids are the potent water-soluble antioxidants and free radical scavengers. They tend to prevent oxidative cell damage, have strong anti-cancer activity as they inhibit the initiation, promotion and progression of tumors. They are also known to reduce the risk of heart diseases. Two important phenolic compounds, Phloretin and Galangin are potentially active against a wide range of bacteria, fungi and viruses.

- ***Phenols and Phenolic acids***

  Free phenols are relatively less common in plants. Hydroquinone is one of the most widely distributed, while catechol, orcinol, phloroglucinol and pyrogallol, have also been reported from a few plant species.

  By contrast to free phenols, phenolic acids are widely spread throughout the plant kingdom. They are also known as hydroxybenzoates, gallic acid being the principal component. Universal among the angiosperms are hydroxybenzoic, protocatechuic, vanillic, syringic, gentisic, cinnamic, *p*-coumaric, caffeic and ferulic and salicylic acids. Phenols and phenolic acids have played a potential protective role against oxidative damage diseases (coronary heart disease, stroke, and cancers). Olive oil phenolic compounds have been reported to possess important biological activities that may exert positive effects on plasma lipoproteins, oxidative damage, inflammatory markers, platelet and cellular function, antimicrobial activity and bone health, which are associated with the development of chronic degenerative diseases. Polyphenols have long been associated with anti-inflammation, anti-aging effects especially in skin and deactivation of oxidant species.

3. Terpenes and Terpenoids

Terpenes or isoprenoids are a large and diverse class of metabolites, produced by a variety of plants. They are the primary constituents of the essential oils of many types of plants and flowers. Terpenes are derived basically from isoprene units, with molecular formula $(C_5H_8)_n$. A wide range of terpenes have been identified such as monoterpenes , sesquiterpenes, diterpenes, triterpenes, tetraterpenes, polyterpenes, hemiterpenes and nor-isoprenoids in foods and essential oils of various plant species. Terpenes have been found to have anti-bacterial, anti-malarial and anti-cancer effects. Gossypol, a diterpenoid is produced by *Gossypium hirsutum* (cotton) has strong antifungal and antibacterial properties. Avenacin, a triterpenoid of *Triticum aestivum* (wheat) possess inhibitory action against *Gaeumannomyces graminis.* Besides, luteolin and friedelin terpenoids obtained from *Anemone blanda* (winter windflower) leaves have potent insecticidal activity. Terpenes have been successfully employed for the treatment of acne (terpineols), acute or chronic bronchitis and related disorders. Due to comparatively low toxicity and high bioavailability of dietary monoterpenes, these compounds have been frequently used as potential therapeutic agents.

4. Tannins

Tannin is a polyphenolic compound which binds and precipitates various organic compounds including proteins, amino acids and alkaloids. They are widely distributed throughout the plant kingdom and are found in almost every plant part including stem bark, wood, leaves, fruits and roots. Tannins have been reported from180 families of dicotyledons and 44 families of monocotyledons. Being secondary metabolites, tannins play a role in protection from predation, as pesticides and probably in growth regulation in plants.

Consumption of tannin-containing beverages, especially green teas and red wines, has been effective to cure variety of health problems. After the isolation of geraniin from *Geranii herba,* recently, various tannin compounds belonging to different classes like oligomeric hydrolyzable tannins or complex tannins have been isolated from various medicinal plants. Apparent biological and pharmacological activities such as inhibition of carcinogenesis, host-mediated antitumor and antiviral activities, inhibition of lipid peroxidation and lipoxygenase, xanthine oxidase, and monoamine oxidase have been associated with several of these tannin compounds.

Besides, some important human physiological processes, including stimulation of phagocytic cells and even a wide range of anti-infective actions, have been attributed to tannins. Tannins have been reported to possess toxic effect on filamentous fungi, yeasts and bacteria. They possibly bind to the bacterial cell walls and prevent their growth and protease activity.

5. Quinones

Quinones are compounds with either a 1,4- diketocyclohexa-2,5-dienoid or 1,2- diketocyclohexa-3,5-dienoid moiety. Broadly speaking, quinones belong to three major classes including benzoquinones, napthoquinones and anthraquinones. They have been experienced in almost all parts and organs of plants. Anthraquinones is the largest group of plant quinones. They are widely distributed in plant kingdom, especially in *Cassia, Aloe, Rheum, Rhamnus, Rubia* and *Tectona.* The most important anthra-quinones isolated from plants include aloe-emodin, chrysophanol, emodin, physcion, cassiamin C, sennoside A and hypericin.

Hypercin, an anthroquinone, obtained from *Hypericum perforatum* (St. John's wort) has been clinically used as an antidepressant, antiviral and antimicrobial agent. Some other well known quinones like Barbaloin and emodin (antiviral and antimicrobial), alizarin, purpurin (antigenotoxic), Quinalizarin, rhein, protohypericin, emodin bianthrone and emodin anthrone (anti-humancytomegalovirus activity), chrysophanic acid (anti-poliovirus types 2 and 3 activity) are potentially active agents.

Additionally, quinones also act as primary antioxidants by forming stabilized radicals, which interrupts the autoxidation chain reaction.

6. Glycosides

Glycosides are the organic compounds comprising a sugar portion linked to a non-sugar moiety (an alcohol, a phenol or a complex molecule) in a particular manner. The molecule from which the non-sugar portion of glycoside is derived is called aglycones or genin. Plants contains a variety of glycosidal substances like anthraquinone glycosides, cardiac glycosides, cyanogenic glycosides, sterol glycosides, glucosinolates glycosidal dyes and pigments etc.

- ***Cardiac glycosides***

  Cardiac glycosides are present in the plants of Scrophulariaceae (figwort family) particularly in Digitalis purpura (foxglove) and in Cardiac glycosides such as digitoxin, digoxin, convallotoxin strengthens heart and controls heart contraction rates. The aglycones of cardiac glycosides have a steroidal structure. They inhibit Na+/K+- ATPase-pumps in the cell membranes. These pumps are critical in the functioning of cardiac cells. Cardiac glycosides are also diuretic in nature and stimulate urine production and aids in the removal of fluid from tissues.

- *Cyanogenic glycosides*

  These glycosides are poisonous and characterised by the presence of cyanide. In small doses they serve as muscle relaxant. The bark of wild cherry and leaves of elderberry (Sambucus racemosa) contain cyanogenic glycosides, which can be used to control and cure dry coughs.

7. **Saponins**

Saponins are the class of glycosides which foam when shaken with water. The foaming ability of saponins is due to the combination of a hydrophobic sapogenin and a hydrophilic sugar part. In plants saponins occur in two main forms: steroidal and triterpenoid saponins. Saponins are common in beans, vegetables and several members of the lily family. The best known source of saponins is peas, soybeans, soapwort, soaproot, soapbark and soapberry. Saponins are associated with reduction of blood cholestrol, lower the risk of human cancers (anti-tumor and anti-mutagenic), immunity booster, reduction of bone loss and free radical scavenger.

Saponins have been excessively used in shampoos and other cosmetic preparations. The pods of *Acacia concinna, Aesculus indica, and Sapindus mukorosii* are being used in the preparation of shampoos. Recently saponins have been found to possess spermicidal, cardiovascular and fungicidal properties.

8. **Essential oils**

Volatile and commonly odorous liquids, based on isoprene units are the essential oils. With some exceptions including camphor, thymol and menthol are solids. These essential oils have bactericidal properties and are widely used as food preservatives. Traditionally some essential oils are also used as local anaesthetic and a source of snuff.

9. **Mucilage and Gums**

These are non-nitrogenous compounds related to carbohydrates. They are soluble with water and are not absorbed by the skin or mucous membranes. Most of the mucilages and gums are obtained as exudates from incisions in the bark of trees. The most important exudate is gum arabic, a product of several species of *Acacia*. This gum is produced by the action of microbe, *Bacterium acacieas*. Other plant species which yield gum are *Anogeissus latifolia, Cyamposis tetragonolobus, Sesbania bispinosa, Cassia* sp., *Mangifera indica* etc.

Mucilage is the characteristic of the members of Malvaceae family. Several species of Mallow (*Malva*) and of marshmallow (*Hibiscus*) yield mucilage from leaves and fruits which serve to soothe irritated mucous membranes of throat and digestive tract. Mucilage obtained from the fruits of *Lallemantia royleana* and the linseed is used in the dressing if wounds and boils.

10. Fixed oils or fats

Seeds are the seat of most of the fats. Fixed oils are called so, because in contrast to volatile or essential oils they do not distil at temperature of boiling water. Chemically they are glycerides of higher saturated or unsaturated fatty acids. Highly unsaturated fixed oils are used as cathartics or mind laxative.

11. Toxalbumins

These are poisonous, irritant proteinaceous substances, commonly found in some seeds. They may cause inflammation when applied to mucous membranes and purgation or emesis when swallowed. Examples of such toxalbumins are the seeds of *Abrus precatorius* (jequrity bean) and *Cassia absus*. Several of these seeds have been used as anthelmintics but due to their high toxicity they are less practiced.

2.3 Some important medicinal herbs: their identification & utilization

2.3.1 *Catharanthus roseus* (L.) G. Don

Catharanthus roseus (L.), commonly known as "sadabahar," is an important medicinal plant of the family Apocynaceae. *C. roseus has* two cultivars distinguished on the basisof their flower colour, one is pink flowered "Rosea" **(Fig. 6)**, while other is white flowered "Alba". In english the plant is known by "Cape Periwinkle", "Rose Periwinkle", "Rosy Periwinkle" and "Old Maid".

Fig. 6: *Catharanthus roseus*

Systemic position

Kingdom	:	Plantae
Division	:	Magnoliophyta
Class	:	Magnoliopsida
Order	:	Gentianales
Family	:	Apocynaceae
Genus	:	*Catharanthus*
Species	:	*roseus*

Synonyms: *Vinca rosea, Ammocallis rosea* and *Lochnera rosea*

Distribution

Catharanthus roseus is native to the Indian Ocean Island of Madagascar. In the wild, it is found to be an endangered plant and the main cause of their decline is the habitat destruction by the slash and burnagriculture however, it is now common in many tropical and subtropical regions worldwide, including the Southern United states. In US, the plant is widespread in South Carolina, Taxas, Mississippi, Florida and Georgia. In India it is mostly found in Tamil Nadu, Karnataka, Gujrat, Orissa, West Bengal and Andhra Pradesh

Identification

- An evergreen, tender, perennial, sub-herb or herbaceous plant growing to 1 metre.
- Presence of milky latex and unpleasent smell.
- Main source of vinca alkaloids, vincristine and vinblastine.
- Presence of narrowly winged stems.
- Leathery, dark green leaves arranged in opposite pairs.
- Flowers white or pink, comprising five petals spreading from a long, tubular throat.
- Fruits made up of two narrow, cylindrical follicles with numerous grooved seeds.
- Seeds small in size and black in colour.

Potentially Active Chemical Constituents

Catharanthus roseus possess carbohydrate, flavinoid, saponins and alkaloids. Alkaloids are the most potentially active chemical constituents of *Catharanthus roseus*. More than 400 alkaloids are present in the plant, which areused as pharmaceuticals, agrochemicals, flavor and fragrance, ingredients, food additives and pesticides.The alkaloids like actineo plastidemeric, Vinblastin, Vincrestine, Vindesine, Vindeline Tabersonine etc.are mainly present in aerial parts whereas ajmalicine, vinceine, vineamine, raubasin, reserpine, catharanthine etc are present in roots and basal stem. Rosindin is an anthocyanin pigment found in theflower of *C. roseus*.

Medicinal Uses (Health Benefits)

Catharanthus roseus has long been used to treat a wide array of diseases in the traditional system of medicine. The plant is highly practiced in the following health ailments.

1. **Cardiotonic**

   A cardiotonic is a substance that has a favorable effect upon the action of heart. C. *roseus* is a versatile medicinal herb known for its anticancer action. *Catahranthus roseus* is known for its cardiotonic agent and have a strengthening effect on the heart.

   An alkaloid Reserpine is obtained from the whole plant of *C.roseus*, which is one of the most effective natural herbal medicines to control hypertension. Reserpine lowers blood pressure by decreasing the levels of certain chemicals like norepinephrine in blood. This allows blood vessels (veins and arteries) to relax (widen) and heart to beat more smoothly and easily.

   A monomeric alkaloid ajmalicine/raubacine found in the roots of *C. roseus* has been found to have a broad application of cardiac diseases, especially in the obstruction of normal cerebral blood flow.

   In cook Island (a nation of south pacific) the decoction of dried leaves of *C. roseus* is taken orally to treat hypertension.

   C.roseus alkoilods acts as anti-haemorrhagic, thereby seals injured blood vessels.

   Catharanthine, an alkaloid isolated from *Catharanthus roseus* dilates small mesenteric arteries and decreases heart rate and cardiac contractility by inhibition of voltage-gated calcium channels on vascular smooth muscle cells and cardiomyocytes.

   C.roseus leaves extract is effective in controlling cardiovascular conditions like heart weight, blood glucose level, serum cholesterol level and serum triglyceride level.

   Alcoholic extract of leaf and stem is administered orally as cardiotonic, to pump enough blood supply to other organs.

2. **Cancer**

   Catharanthus roseus is known for the production of anticancerous drugs. The important anticancer alkaloids, vinblastine and vincristineare derived from stem and leaf of *Catharanthus roseus*. These alkaloids have growth inhibition effecton some human tumors. Vinblastine is recommended in the treatment of neoplasms, Hodgkin's disease while Vincristine is used to cure leukemia in children.

3. **Diabetes**

   Extracts from leaves and flowers of C. *roseus* reduces sugar level in the blood. The Hypo glycemic effect in C. *roseus* is due to an increase of glucose utilization in the liver.The hot water extract of dried whole plant is taken orally to reduce diabetes mellitus.

4. **Brain health**

   Daily supplements produced from the active ingredientsof C. roseus helps to facilitate optimum blood supply to the brain. It also promotes the level of oxygen and glucose required for the healthy functioning of the brain. It increases the levels of serotonin (a compound found in serum and blood platelets that contracts the blood vessels and act as a neurotransmitter) in the brain.

5. **Other miscellaneous uses**
   - It is antiviral and antibacterial in nature, therefore, it potentially work against viral and bacterial infections.
   - *Catharanthus roseus* extracts are potentially used to counter dysentery.
   - It has the anti-inflammatory action which heals the inflammation.
   - It is a skin toning herb that promotes glow and rejuvenates the skin. Thus, it effectively cures skin diseases.
   - It is a potent natural herb to reduce acne. It also relieves eczema and suppresses the swelling of the skin.
   - Treats bleeding of nose, often usedto prevent gingivitis.
   - It protects the eyes from inflammation, effectively used to cure conjunctivitis.
   - It is widely used to cure aphthous ulcers.
   - It is a helpful remedy for headache. During fatigue, it relaxes the mind and body and helps to gain the energy back.
   - *Catharanthus roseus* counters depression and nausea.
   - It eases the breathing problem in asthmatic patients.
   - It is used to suppress swelling and works against insect bites.
   - It guards against urinary problems.

2.3.2 *Centella asiatica* (L.) urban

Centella asiatica is a clonal, perennial herbaceous creeper **(Fig. 7),** belonging to the family *Umbellifere* (*Apiceae*). The plant is commonly known as **gotu kola** or **Mandukparni.** In English, it is known as water navel, water pennywort and Indian pennywort. It is one of the reported "miracle elixirs of life" known over 2000 years ago. It is an ancient Indian herb found throughout India.

Fig. 7: ***Centella asiatica***

Distribution

The plant is indigenous to the warmer regions of both the hemispheres, including Asia, Africa, Australia, southern United States of America, Central America and South America. It is especially abundant in the swampy areas of India, up to an altitude of approximately 700 m.

Systematic Position

Kingdom	:	Plantae
Subkingdom	:	Tracheobionta
Superdivision	:	Spermatophyta
Division	:	Magnoliophyta
Class	:	Magnoliopsida
Subclass	:	Rosidae
Order	:	Apiales
Family	:	Apiaceae
Genus	:	*Centella*
Species	:	*asiatica*

Synonyms

Hydrocotyle asiatica L.

Trisanthuscochin chinensis

Vernacular names

This plant is known by an array of vernacular names in different regions of the globe **(Table 4).**

Table 4: Different vernacular names of *Centella asiatica* used in different languages

Language	Vernacular Name
Chinese	LueiGongGen, Tungchian
English	Gotukola,IndianPennywort
French	Hydrocotyleasiatique
German	AsiatischerWassernabel
Gujrati	Khandbhrami
Hindi	Bengsaag
Italian	Idrocotile
Japanese	Tsubokusa
Malay	Pegaga
Mauritius	Bavilacqua
Sanskrit	Mandukparni Gotukola
Sinhalese	Vallarai
Tamil	Kodangal

Identification

The plant species is identified on the basis of following charactersitics:

1. The plant is a small trailing herb.
2. Stem is glabrous, pink to purplish, striated and rooting at nodes.
3. Leaves are fleshy, orbicular to reniform and dentate, solitary or in groups of upto five.
4. Petiole is long, smooth on upper surface and hairy below.
5. Flowers are pink and white in fascicled umbels. Flowers are 2 to 8, inconspicous with an involurce of 2-ovates, membranous persistent bracts.
6. The fruits are oblong, dull brown, laterally compressed, pericarp hard, thickened and woody white.
7. Rosettes of calcium oxalate are found within the palisade cells.
8. Epidermis of the leaf has striated cuticle and numerous stomata.

Chemical Composition

The plant contains more than 70 phytoconstituents including triterpenic acid, triterpenoid saponins, polyacetylenes, flavones, sterols, and lipids.The major bioactive compounds responsible for its medicinal effects are asiaticoside, asiatic acid and madecassic acid. The triterpenic steroids reported to occur in C.*asiatica* are stigmasterol and sitosterol. Major constituents of plant leaves indicate that they contain 89.3% water, 6.9% carbohydrate, 2% fibre, 1.6% protein, and 0.6% fat. 100 g of leaves contains approximately 1.6% ash, 414 mg K, 170 mg Ca, 30 mg P, 3.1 mg Fe, 6.58 mg carotene, 0.15 mg thiamine, 0.14 mg riboflavin, 1.2 mg niacin, and 4 mg ascorbic acid. The most common active compounds identified in gotu kola are presented in **Table 5**.

Table 5: Phytoconstituents in *Centella asiatica*

Phytocompound Class	Phytoconstituent
Triterpenic acids	Asiatic acid
	Madasiatic acid
	Madecassic acid
	Thankunicacid
	Indocentoic acid
	Euscaphic acid
	Terminolicacid
	Isothankunic acid
Isothankunic acid Sugar esters	Asiaticoside
	AsiaticosideA, B, C, D, E, F
	Braminoside
	Brahmoside
	Brahminoside

Phytocompound Class	Phytoconstituent
Triterpenic steroids	Stigmasterol
	Sitosterol
Essentialoil	Terpeneacetate
	Pinene
	Caryophyllene
Flavone derivatives	Quercetinglycoside
	Kaempferol
	Astragalin
	Catechin
	Rutin
	Naringin
Sesquiterpenes	Elemene
	Bicycloelemene
	Trans-farnesene
	Ermacrene D

Medicinal Utilization (Health benefits)

Centella asiatica has been used for medicinal purposes since prehistoric times. C. asiatica has been considered by Vagbhatt as the best herb for improving memory and intellect. It belongs to the group of drugs known as medhyarasayanas (psychotropic drugs). Locals as well as traditional healer's from Indonesia, Madagascar, India, Africa and other parts of the globe utilize this medicinal herb both topically and orally for various routine ailments. This plant was officially included in the Indian Pharma in the19th century. Some of the major health benefits of this potent medicinal plant species are discussed under:

1. **Memory enhancer and brain health**

   Whole plant is highly practised for the enhancement of memory of patients with weak memory. Mentally retarded children fed with *Centella asiatica* improve general menta labilities.*Centella asiatica* promotes the release of brain derived neurotropic factor, a protein that acts like fertilizer for brain to encourage new brain cell formation. *Centella asiatica* increases nerve growth factor which stimulates the growth of new cells and also promotes the safeguards of the existing nerve cells. *Centella asiatica* when taken together with vitamin E and C improves memory, thus making *Centella asiatica* another powerful brain memory enhancer. *Centella asiatica* is helpful in the treatment of Alzheimer's disease (patients have significantly reduced levels of acetylcholine, a neurotransmitter responsible for memory and learning).

   Centella asiatica improves ones mood, increases calmness, contentedness, and alertness and reduces symptoms of anxiety and depression.

Because of its great significance as memory enhancer and brain tonic in ayurvedic system of medicine, it is regarded as a brain food. It helps to energies brain and nerve cells. The leaves of this plant are taken readily for the development of brain. Powdered leaves along with cow's milk are recommended in mental weakness and memory loss. The mixture of gotu kola, brahmi, ginkgo, cat's claw, and rosemary helps in cognitive development in both children and adults. It acts upon the circulatory system and thereby more oxygen flows to the brain and therefore promotes cognition. The free radical scavenging potential of this plant ensures breaking of plaque in the brain and thus tends to prevent degeneration of neural structures.

C. asiatica herb is an effective natural remedy to insomnia and sleeplessness. Powder of *C. asiatica* (3gm) is mixed with cow's milk and taken orally for better efficacy. The soothing feature of it reduces stress and stimulates sound sleep. Along with other ayurvedic preparations, the *C. asiatica* is an efficient tonic for nerves and the nervous system as a whole. The herb is mixed with other Ayurvedic medicinal formulations and taken empty stomach early in the morning for the treatment of epilepsy (seizure disorder).

2. **Skin care**

Since the ancient times, this medicinal herb is used to treat many skin related disorders such as eczema, erysipelas, leprosy, psoriasis etc. The powder of gotu kola and vaselinis externally applied on the affected areas of skin both in the morning and evening. The application of its extract helps to reduce skin scars. In herbal system of medicine, cream is prepared from this herb, which helps to reduce scarring. It is helpful to minimize pregnancy marks when mixed with vitamin E.

3. **Hair care**

C. asiatica is used topicallyfor the growth of healthy hair, by acting on the hair follicle region. It helps to stop hair loss and prevents it from graying.In equal proportion, gotu kola juice and coconut oil is mixed and cooled. The mixture is massagedon the scalp and enhances hair health by preventing it from dryness and combat stress caused by free radicals.

4. **Respiratory disorders**

The herb along with cow's milk is readily used in the treatment of respiratory problems such as nasal congestion, respiratory blockage, bronchitis and asthma.

5. **Female sterility**

Dried powder of *Centella asiatica* and *Trichodesma indicum* mixture (2:1 ratio) is taken thrice a day just after menses to treat sterility.

6. **Ageing associated disorders**

Gotu kola is a natural anti-aging adaptogenicagent. It reduces the effects of physical and mental stress on the health of the person. It creates a restorative and rejuvenating influence on the nervous system, which helps to prevent disorders arising due to a stressful lifestyle such as diabetes, and hypertension.The anti-aging property of this herb helps in delaying the degeneration of the tissues in the body. It helps to protect the bones, joints, skin, brain, and the nerves from the effect of age-related wear and tear.

7. **Other medicinal uses**
   - *Centella asiatica* allows blood flow smoothly in the body, increases the count of white blood cells and purifies the blood.
   - *Centella asiaticais* used as a rejuvenating agent.
   - Because of its diuretic nature, it is used for the treatment of dysuria.
   - As a tranquilizing agent, it is used for the treatment of hysteria, agitation, insomnia and epilepsy.
   - Used to cure urinary tract disorders.
   - Found to be beneficial for healing the wounds.
   - Gotu kola enhances the immunity of the body.
   - Used to reduce high blood pressure (hypertension).
   - Cures arthritis and joint pain.
   - Cures burns, removes the scars of injury or chicken pox.
   - Helpful in the treatment of liver and kidney disorders.
   - In the treatment of gastrointestinal problems.
   - Brings down the body temperature caused due to fever.
   - Useful to children in bowel disorders.
   - In Thailand, gotu kola is used to detoxify the Opium poisoning.
   - Used in the treatment of Jaundice and bleeding disorders.
   - Paste of the whole plant is applied against elephantiasis.

2.3.3 *Dioscorea deltoidea Wall. ex Griseb.*

Dioscorea deltoidea, commonly known as "*yam,* is a climbing herb with rhizomatous root stock **(Fig. 8).** Underground rhizome of this plant is commonly known as *wild yam* or *Singli-mingli* in hindi and *Varahikand* in sanskrit. It belongs to class spermatopsida and family Dioscoreaceae. *Dioscorea deltoidea* was named after the ancient Greek physician and botanist, Dioscorioids. The genus *Dioscorea* belongs to monocotyledons and comprises 350-400 species, distributed throughout the tropics and subtropical regions in the world, mainly in West Africa, parts of Central America and the Caribbean, The Pacific islands and South East Asia. *Dioscorea deltoidea* is mainly found in India, China, Nepal, Bhutan, Pakistan, Afghanistan and Vietnam. In India the plant species are mostly distributed in Kashmir to Assam at altitudes of 550-3100m. In IHR (Indian Himalayan regions) it is found in Arunachal Pradesh, Sikkim, Assam, Meghalaya, Jammu & Kashmir, Himanchal Pradesh and Uttrakhand.

Fig. 8: *Dioscorea deltoidea*

Systematic Position

Kingdom	:	Plantae
Division	:	Magnoliophyta
Class	:	Liliopsida
Order	:	Liliales
Family	:	Dioscoreaceae
Genus	:	*Dioscorea*
Species	:	*deltoidea*

Identification

- Hairless vine, twining clockwise
- Ligneous, irregular tubers
- Simple and alternately arranged, 5-11cm long, broad, ovate, often heart shaped leaves
- Male flower spikes solitary in leaf axils
- Flowers in small distant clusters, stamens 6, antheriferous.
- Female spikes solitary, few-flowered
- Capsule 2 cm long, abovate or obcordate
- Seeds winged, unequally all round
- Lignification in phloem
- Epidermis if present, consists of layer of collapsed cells
- Idioblasts contaning bundles of raphide crystals of calcium oxalate in cortex

Chemical Composition

Diosgenin, an aglycone is a chemical substance found in *Dioscorea* and is used commercially in pharmaceutical industry. Apart from diosgenin, dioscorin, dioscin and other alkaloids are also found. Root contains phytosterols, alkaloids, tannin and rich source of starch. Other substance found are aluminium, ascorbic acid, ash, beta-carotene, calcium, chromium, cobalt, iron, magnesium, manganese, niacin, potassium, phosphorus, protein, riboflavin, selenium, silicon, sodium, thiamine, tin, zinc.

Medicinal use of *Dioscorea deltoidea*

Traditionally the plant has been used as an anti-rheumatic and anti-ophthalmic. In the Western India, it has been used as a source of steroid drugs. In Kumaun region of Uttarakhand, rhizome powder of the plant is taken orally to cure dysentery, abdominal pain and piles. In Garhwal region *Dioscorea deltoidea* rhizomes are cut in smaller pieces and made into vegetable or soup (*Saag*) on the holy occasion of Maha-Shivratri. Sometimes the powder mixed with milk is used as a general tonic to improve general health. Rhizomes are also made into soap and used to kill lice especially among hill tribe communities in Jammu & Kashmir, India. In Nepal, auxiliary rhizomes from the upper part of the plant are boiled and the liquid is used to treat gastric problems and dysentery.

Dioscorea deltodia in rheumatism

Dioscorea deltodia is highly practiced for the treatment of rheumatic conditions. In rheumatism the plant is useful in the treatment of the following conditions.

- **Rheumatoid arthritis**

  D. deltodia has anti-inflammatory properties and treat rheumatoid arthritis, it is a disease that causes inflammation of the joints and can cause stiffness that restricts mobility in people with this condition.

- **Cramps and muscular pain**

  As an anti-spasmodic it reduces muscle spasms and muscular pain.

- **Diverticulitis (Diverticula Inflammation)**

  D. deltodia contains many alkaloids which are muscular relaxants that especially target muscles in the abdomen and pelvis. Further, it can improve menstrual cramps and the pain associated with endometriosis.

- *D. deltodia* contains a steroid, diosgenin that has anti-inflammatory effect in animals and proven to reduce inflammation of joints. It causes an inhibition of growth of fibroblast-like synoviocytes from rheumatoid arthritis. Diosgenin acts as anti-osteoarthritis which causes progressive reduction in articular cartilage and synovial inflammation.
- *D. deltodia* contains active ingredients known as steroidal saponins which play a role in alleviating muscle strains, chronic muscle pain and arthritis.
- The roots of *D. deltodia* produce cortisone, a steroidal hormone, which is used in the treatment of rheumatic diseases.

Other medicinal uses

- Due to the presence of poisonous saponins in the root, soap is prepared and used externally as a body wash to kill lice.
- Tubers are boiled with arums, mushroom and made into paste. The paste is applied topically to treat skin diseases. Tuber paste is applied on affected parts to treat peeling feet.
- Mature tubers are boiled and eaten to gain body weight. They are also used in stomachache, constipation, indigestion, abdominal pain.
- Powder of tuber is given orally in abdominal pain after delivery.
- Roasted tubers are taken in leprosy, asthma, cough, cold and tuberculosis.
- Root powder is used as one of the component in herbal preparations to maintain kidney functions.
- Fresh tuber decoction cures laryngitis in children, insect bite, ringworm, goitre and common fever.
- Regulates female sex hormones and is considered as natural herbal remedy to reduce symptoms associated with menopause. Root is used to sooth dysmenorrhoea, reduces uterine and ovarian pain. It is also helpful in relieving labour pain and prevents early miscarriage. Root decoction relives the pain of child birth, good for nausea that can be experienced during pregnancy.
- The herb is used as a tonic in digestive disorders including gall bladder inflammation, irritable bowel syndrome (IBS) etc.
- The plant is used as refrigerant to reduce body heat during summer.
- Tubers are boiled or baked in open fire and eaten to reduce blood cholesterol level.

2.3.4 *Withania somnifera* (L.) Dunal

Withania somnifera is commonly known as ashwagandha, Indian ginseng or winter cherry. The name Ashwagandha is composed of two Sanskrit words, "Ashwa" means *horse* and "Gandha" means *smell*. Because of the strong aroma of the root, it is often described as horse like.It is cosmopolitan and grows throughout the drier parts and subtropical regions. The plant is also reported from India, Pakistan, Afghanistan, Philistine, Egypt, Jordan, Morocco, Sri Lanka, Spain, CanaryIsland, Eastern Africa, Congo, Madagascar and Souths Africa. In India, it is widely distributed in the warmer parts and tropical areas. From north-western region, Maharashtra,Gujarat, Rajasthan, Madhya Pradesh, Orrisa, Uttar Pradesh, Punjab plain it extends to the mountain regions of Punjab, Himachal Pradesh and Jammu. It is a short (35–75 cm), tender, perennial shrub with tomentose branches **(Fig. 9A)**; stout, tuberous, fleshy and whitish-brown roots **(Fig. 9B)**; dull green and elliptic leaves **(Fig. 9D)**; small, green and bell-shaped flowers **(Fig. 9C)** and orange-red fruits **(Fig. 9E)**.

Fig. 9 (A-E): ***Withania somnifera:*** **(A) Whole plant (B) roots (C) flowers (D) leaves (E) fruits**

Systematic Position

Kingdom	:	Plantae
Division	:	Angiosperma
Class	:	Dicotyledons
Order	:	Tubiflorae
Family	:	Solanaceae
Genus	:	*Withania*
Species	:	*somnifera*

Synonyms : *Physalis somnifera*

Vernacular Names

Language		**Vernacular Name**
Hindi	:	Asgandh
Sanskrit	:	Ashvagandha, Balada, Gandhpatri, Kamrupini, Vajini
Gujarati	:	Asam, Asoda, Ghodasoda
Punjabi	:	Asgand
Telugu	:	Asvagandhi, Pennereu, Vajigandha
Tamil	:	Asuragandhi, Amukkira
Bengali	:	Ashvagandh
Urdu	:	Asgandanagaori
Marathi	:	Asgundh, Kanchuki, Askandha
Canarese	:	Amangura, Hirimaddina-gadde, Sogada-bery

Identification

- Almost the complete plant is covered and surrounded with short, fine and branched hairs and silver-grey in colour.
- The roots have a short and uneven fracture, a strong odour and mucilaginous biter and acrid taste.
- Stems are errect and brownish dark in colour.
- Leaves are simple, ovate, glabrous, and up to 10 cm long.
- Flowers are greenish or lurid yellow, small about 1 cm long; few flowers (usually 5) born together in axillary, umbellate cymes (short axillary clusters).
- Fruits are globose berries, 6 mm in diameter, orange red when mature, enclosed in the inflated and membranous persistent calyx.
- Seeds are yellow, discoid, reniform and 2.5 mm in diameter.
- The number of chromosomes is 2n=48.

Chemical Constituents

Withania somnifera is a reservoir of diverse active chemicals in the different parts of the plant which are as follows:

Root

Roots are reported to contain alkaloids, amino acids, steroids, volatile oil, starch, reducing sugars, glycosides, hentriacontane, dulcitol, withaniol. Many biochemically heterogeneous alkaloids have been reported in the roots. Basic alkaloids include cuscohygrine, anahygrine, tropine, pseudotropine, anaferine, isopelletierine, withananine, withananinine, pseudo-withanine, somnine, somniferine, somniferinine. Other alkaloids include withanine, withasomnine, and visamine. The free amino acids identified in the rootsincludeaspartic acid, glycine, tyrosine, alanine, proline, tryptophan, glutamic acid, and cysteine.

Shoot

The tender shoots are rich in crude protein, calcium and phosphorous. They are reported to contain *scopoletin*.

Stem

The stem of the plant contains condensed tannins and flavonoids.

Bark

The bark contains a number of free amino acids.

Leaves

The leaves of the plant are reported to contain 12 withanolides, many free amino acids, chlorogenic acid, glycosides, glucose, condensed tannins, and flavonoids. The leaves of the plant from different habitats contain different withanolides–a group of C_{28} steroids characterized by a 6-membered lactone ring in the 9-carbon atom side chain. Withaferin A, a steroidal lactone is the most important withanolide isolated from the extract of the leaves.

Fruits

The green berries contain amino acids, a proteolytic enzyme, condensed tannins, and flavonoids. They contain a high proportion of free amino acids which include proline, valine, tyrosine, alanine, glycine, hydroxyproline, aspartic acid, glutamic acid and cysteine.

Medicinal Importance

Historically this plant was traditionally used as a rejuvenative herb across the ancient Mediterranean countries. Despite its name, as *the smell of a horse,* ashwagandha offers a lot more benefits than those referring to horse-like attributes of physical strength and virility. Historically ashwagandha has been practiced in Ayurveda, Unani and Middle Eastern traditional systems of medicine.

Central Nervous System Disorders

Withania somnifera is an important medicinal herb to treat diseases from normal infection to cancer but has been effectively used in nervous system diseases. The plant has long been practiced in the treatment of the following CNS disorders

Parkinson's and Alzheimer's Diseases

Parkinson's disease is most common neurodegenerative disorder in elderly patients. It is a slow progressing degenerative disorder of the central nervous system that affects movement, muscle control and balance. It is caused by the death of dopamine-producing cells. While Alzheimers disesase is the most common type of dementia, where the patients usually above 65 years , progressively losses memory and has behavioural issues, due to demage of brain cells.

Withania somnifera preparations combat the Parkinson's and Alzheimer's diseases by the following modes:

- Reduces cell demage and inflammation by having steroids which fights free radicals and increases the activity of natural antioxidants in the body.
- Improves behavoiur related symptyoms in Alzhemer,s disease
- Improves Parkinson's symptoms related to muscle control movements and balance by increasing dopamine levels.
- Ability to repair demaged neurons and reconstruct neuronal network and synapses, which helps to improve memory and cognition.
- Increases the number of certain type of proteins that helps clear out beta-amyloid proteins from the brain, enabling the reversal of Alzhimer's disease.
- Powder of Ashwagandha roots along with *Mucuna pruriens* and *Hyoscyamus reticulatus* seed powder is an effective formulation for the treatment of Parkinson's disease due to its potential anti-oxidant, anti-peroxidative and free radical quenching properties.
- Root extract of *W. somnifera* shows potential neuroprotective effects against H_2O_2 and β- Amyloid cytotoxicity during Alzheimer's disease.
- Dried root extract of *W. somnifera* enhances liver LRP (low density lipoprotein receptor- related protein) and decreases β-Amyloid formation

by Aβ- degrading protease neprilysin (NEP) in brain.

- Active principles from Ashwagandha, like withanolide A, withanoside IV, and withanoside VI improves memory impairment, neurite atrophy, and synaptic loss in the cerebral cortex by restoring pre-synapses and post-synapses both in axons and dendrites in cortical neurons.

Huntington's disease

It is an inherited disorder that causes the progressive degeneration of nerve cells in the brain. It affects ones functional abilities and usually results in movement, thinking and psychiatric disorders. *W. Somnifera* root extract repairs the major imbalance in GABAergic system (affected neurotransmitter GABA) and improves cognitive function, acetylcholinesterase enzyme activity and glutathione enzyme level in HD patients. Treatment with *W. somnifera* restores impaired motor function and other cognitive deficits. The antioxidant property of the W. somnifera root extract makes it a potential leader to treat HD.

Epilepsy

A disorder in which nerve cell activity in the brain is disturbed, causing seizures.

W. somnifera restores imbalance in GABA/glutamate modulation and higher serum level of peptide hormone, Ghrelin. *W. somnifera* as a complementary and alternative medicine in epilepsy treatment increases GABA level & cortical muscarinic ACh and enhances neurite regeneration in brain, by antioxidative mechanisms. This affordable and culturally acceptable herbal remedy may open new doors for epileptic patients around the world.

Anxiety& Depression

A widespread psychopathological disorder, associated with unpleasant emotional state leads to depression, somatic distress and low self-esteem. Ashwagandha is best known for its ability to reduce stress. It blocks the stress pathway in the brain by regulating chemical signaling in the nervous system. A steroid, withanolides from W. somnifera roots induces GABA-mimetic activity in the treatment of anxiety.

Administration of 600 mg of ashwagandha per day reduces approximately 70-80% severe depression in patients.

OCD (Obssesive Compusive Disorder)

A mental disorder characterized by persistent and distressed thought (Obsessions) with repeated egoistic behavior (compulsion). Ashwagandha root is a safe and effective adjunct therapy with SSRI s (Selective Serotonin Re-uptake Inhibitors) and decreases the severity of symptoms of patients with OCD. Due to its anxiolytic & anti-depressent properties, it is highly recommended in the OCD treatment. Root extracts of Ashwagandha increases serotonergic transmission by affecting 5HT2A/2C receptors in the brain.

Hypoxia

Hypoxia is considered to generate superoxide radicals because of less availability of oxygen which causes neurodegeneration and cognitive dysfunction including memory impairment. Withanolide A, a major component of *withania* root prevents neurodegeneration by modulating hippocampal glutathione biosynthesis during hypoxia. It maintains balance of corticosterone level by increasing minerlocorticoid receptor expression and decreases glucocorticoid expression in hippocampus, which plays a vital role in memory and neuronal survivability.

Catalepsy

A medical condition characterized by a trance or sezure with a loss of sensation and consciousness acompained by rigidity of the body. *W. somniferat* in combination with *Ocimum sanctum, Camellia sinensis,* triphala and shilajit plays an important role in the treatment of Neuroleptic catalepsy by reversing D1 &D2 receptor agonists. Ashwagandha root extract has anticataleptic effect on haloperoidol-induced catalepsy in both acute and chronic conditions. It is believed that antioxidant potential of ashwagandha root contributes to the anticataleptic effect.

Bipolar Disorder (manic depression)

Bipolar disorder is a mental illness marked by extreme shifts in mood, associated with anxiety and insomnia. Ashwagandha extracts appear to improve auditory-verbal working memory (digit span backward), a measure of reaction time and social cognition in bipolar disorder.

Limited treatment availability makes herbal medication an attractive tool against bipolar conditions. Ashwagandha considerably reduces insomnic and anxiety conditions. Daily consumption of standardized ashwagandha root and leaf extracts may improve several areasc of cognitive effect in in bipolar disorder patients.

Additionally the different parts of Ashwagandha are utilized in various parts of the globe for the treatment of following routine human maladies:

- Aging and longevity
- Arthritis and other inflammatory conditions
- Asthma
- Gastrointestinal disorders
- Controls blood sugar levels
- Fights cancer and tends to reduce the side effects of chemotherapy without interfering with the tumor cell-killing activity.
- Ashwagandha roots strengthen the heart muscles and controls cholesterol levels.
- Increases fertility in men by increasingtestosterone levels, improves semen quality, sperm count and sperm mobility.
- Improves the muscular strength of the lower limbs and helps to recover weakness
- Roots stimulates the secretion of thyroid hormones during hypothyroidism
- Ashwagandha capsules are helpful in increasing the red blood cells, white blood cells, and platelet count, which in turn help boost immunity.

- Effective in controlling bacterial infections in humans, effective in urinogenital, gastrointestinal and respiratory tract infections when consumed orally.
- Increases and thickens nutritive value of milk in nursing mothers.
- Ashwagandha powder mixed with oil reduces and cures cutaneous infections.

UNIT - III

Pharmacognosy

3.1 Pharmacognosy: Meaning and definition

The term Pharmacognosy, initially known as *"materia medica"* comes from two Greek words; "**Pharmakon**" meaning drug or medicine and "**gnosis**" meaning knowledge. Pharmacognosy, therefore, is the knowledge of drugs or medicines. The term was first coined by C.A. Sedler in his dissertation , *Analecta Pharmacognostica,* published in 1815 and was referred as the father of Pharmacognosy.

During the 19th century, pharmacognosy was used to define the branch of medicine or commodity sciences which deals with drugs in their crude form. However in the beginning of the 20th century, the subject had developed mainly on the botanical side, particularly concerned with the description and identification of drugs.

Pharmacognosy is an interdisciplinary science which covers all the aspects of drugs of natural origin. The **American Society of Pharmacognosy** defined it as the study of physical, chemical, biochemical and biological properties of drugs, drug substances or potential drugs or drug substances of natural origin as well as the search for new drugs from natural sources. Mostly pharmacognostic studies focus on plants and medicines derived from plants but recently other types of organisms particularly microbes (bacteria, fungi), and several marine organisms are also regarded as pharmacognostically interesting.

3.2 Historical development

The history of herbal medication is as recent as human civilization. Herbal medicines as the major remedy in ancient system of medicine are employed in medical practices since antiquity.

The early medicines of Pharaohs (3000 BC), the Greek (460–370 BC; Hippocratis), the Roman (37 BC; Disoscorides, a Greek physician of the first century AD was the writer of the first Materia Medica (78 AD). They described 600 medicinal plants and those of Middle Ages exemplified by the Arab Physicians (Rhazes 865–925; Avicenna 980–1037) relied mainly on plants for therapy.

India has renowned for practicing classical medicinal systems such as: Siddha, Buddha, Ayurveda and Unani methods of medication and treatment. These medicinal systems are found even in the ancient Vedas and other ancient literatures and scriptures. In India, the Ayurveda concept appeared and grew up between 500 and 2500 BC. The authentic meaning of Ayurveda is "science of life," because ancient Indian system of health care focused on views of human and their sickness. It has been pointed out that the positive health means metabolically well-balanced human beings.

3.3 Modern concept

Medicinal plants have a vital role in the development of new drugs **(Table 6)**. During the years 1950–1970, nearly hundreds of new plant-based drugs were introduced into the USA drug markets, including ricinin, derbipidine, phenplastin, phenicristine etc. From 1971 to 1990, a substantial number of new drugs, such as octoposide, teneboside, *E*- and *Z*-guggulsterone, nebulon, plonotol, artemisinin etc. appeared all over the world. The chemical structure of some of the vital modern phytochemicals is depicted in **Figure 10.**

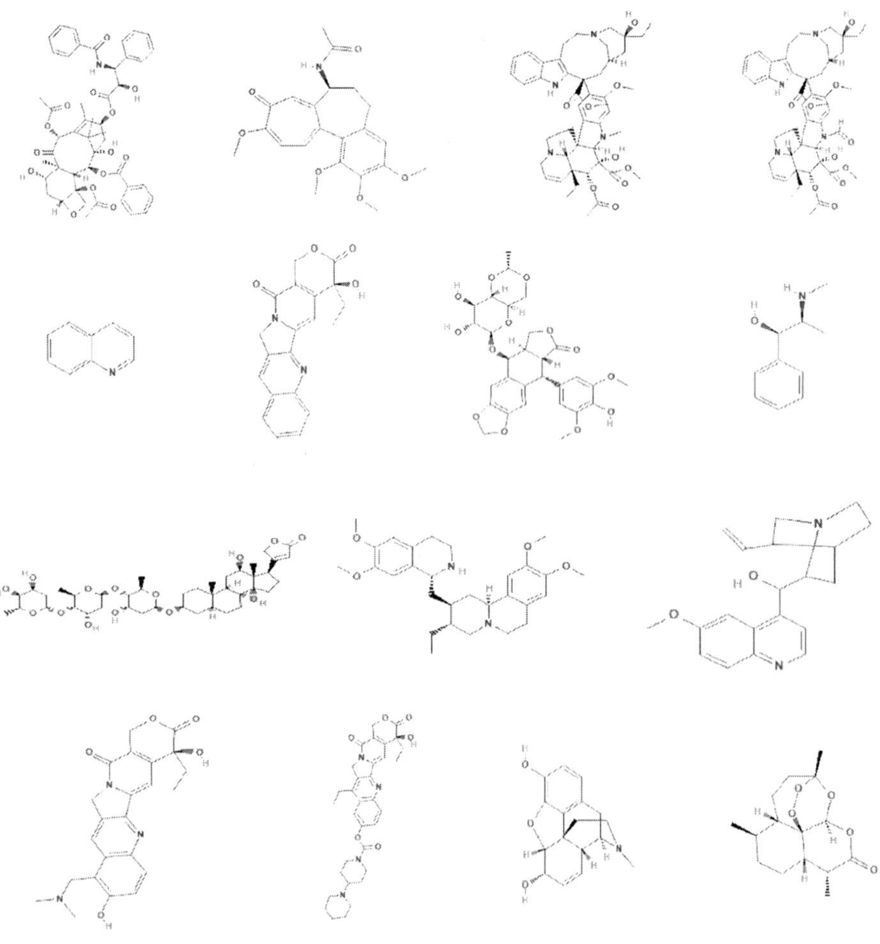

(Taxol, Colchicine, Vinblastine, Vincristine, Quinoline, Camptothecin, Etoposide, Ephedrine, Digoxin, Emetine, Quinine, Topotecan, Irinotecan, Morphine, Artemisinin)

Fig. 10: Chemical structure of some vital plant based modern drugs

About 2% of drugs were introduced from 1991 to 1995 including paciltaxel, toptecan, irinotecan (approved drug, FDA, USA), etc. Plant-based drugs provide outstanding contribution to modern therapeutics such as Vinblastine, isolated from the *Catharanthus rosesus* is used for the treatment of nature preceding: Hodgkin's chorio carcinoma, non-Hodgkin's lymphomas, leukemia in children, testicular, and neck cancer. Vincristine is recommended for acute lymphocytic leukemia in childhood advanced stages of Hodgkin's lymophoma, small cell lung, cervical, and breast cancer. Podophyllotoxin, isolated from *Podophyllum emodi,* is used against small lung cancer cells and lymphomas (testicular cancer). Two compounds (etoposide and teniposide) isolated from one of the *Podophyllum* species are used for the treatment of testicular and lung cancer. Taxol, a well known secondary metabolite from *Taxus brevifolius* is used for the treatment of lung cancer and ovarian cancer. Harringtonine and homo-harringtonine isolated from *Cephalotaxus fortunei* has been used in China for the treatment of acute chronic myelogenous leukemia and myelogenous leukemia. Elliptinium, a *N*-methyl derivative of ellipticine isolated from *Bleekeria vitensis* is marketed in different places in Europe (France) for the breast cancer treatment.

More than hundred plants have been found to possess notable antibacterial and antidiabetic properties. The above-mentioned drugs came into use through the screening analysis of different medicinal plants having lesser side effects and toxicity.

Table 6: Important modern drugs and their plant source

Drug	Plant Source	Modern Application
Vinblastine and Vincristine	*Catharanthus rosesus*	Hodgkin's, Lymphoma, children leukemia, bladder, brain and melanoma
Podophyllotoxin	*Podophyllum emodi*	Testicular cancer, lymphoma and lung cancer
Paciltaxel and Taxotere	*Taxus brevifolia*	Ovarian, malignant melanoma, breast,lungs and pancreatic cancers
Camptothecin, Irinotecan and Topotecan	*Mappia foetida*	Lung, cervical, breast and ovarian cancer
Quinoline and Camptothecin	*Camptotheca acuminata*	Skin, lungs and cervical cancer
Teniposide and Etoposide	*Juniperus communis*	Lung cancer and lymphocytic leukemia
Digoxin	*Digitalis lanata*	Heart therapy
Emetine	*Cephaelis sp*	Induces vomiting and cures dysentery
Ephedrine	*Ephedra sinica*	Respiratory ailments
Quinine	*Cinchona sp*	Treatment of Malaria
Colchicine	*Colchicum autumnale*	Treatment of gout
Artemisinin	*Artemisia annua*	Treatment of Malaria

3.4 Scope of pharmacognosy

Pharmacognosy has been a field of multidisciplinary science and phytochemistry, phytomedicine and phytopharmacology have become important parts of this field.

Molecular biology field has become an important area for medicinal plant drug discovery through the application of screening assays and modern pharmacognosy encapsulates all of these relevant new research areas into a distinct interdisciplinary natural product science.

The focus of research in pharmacognosy have alternated significantly, focusing on isolation and structure elucidation of drugs to their biological activities as well as structure activity relationship studies. Advanced researches in the field of ethnobiomedicine, ethnobotany and ethnopharmacology has also become an essential element in the orbit of pharmacognosy. Pharmacognosy forms an important association between medicinal chemistry and pharmacological studies (pharmaceutical chemistry)

In recent years, as a result of development in phytochemistry and pharmacological testing methods, new plant-derived drugs are finding their way into medicine as single phytochemical, rather than in the mixture form of traditional herbal preparations. The world is now moving towards the herbal medicine or phytomedicines that repair and strengthens bodily systems.

However, presently, drug discoveries are increasing rapidly after adopting traditional/ folk medicine-based uses/approaches to increase results and with safety concerns. Thus, different sub-branches of pharmacognosy, such as analytical, industrial and clinical, have been established as a modern and professional off shoots of specialized pharmacognosy to meet the most productive advancements and collaborations in this field. Furthermore, molecular, metabolomic and genomic pharmacognosy have been introduced as the new and promising targets of research for accommodating future supply and demands in biomedicine, molecular biology, biotechnology, and analytical chemistry of traditional natural medicines and folk medicinal plants.

Pharmacognosy is an important link between pharmaceuticals and basic science as well as ayurvedic and allopathic system of medicines. So pharmacognosy is a science of active principles of crude drugs and can be helpful in dispensing, formulating, and manufacturing of dosage forms. Nevertheless, interdisciplinary combined and collaborative research work is very essential for optimizing the development of traditional biomedicines and pharmacognosy field of research, education and techniques.

3.5 Pharmacognostical scheme

The area of pharmacognosy that studies the composition, usage and history of drugs is known as "Pharmacognostics" and the description of drugs in a systematic manner is known as pharmacognostical scheme, which include the following:

1. **Biological Source**

   This includes the biological names of plants or animals yielding the drug and family to which they belong. Botanical name includes genus and species. Often some abbreviations are written at the end of the botanical name, which represents the author citation e.g *Acacia arabica* Willd. Here Willd indicates the botanist responsible for nomenclature. According to the binomial nomenclature, the botanical name of any plant or animal is always written in italic form, and the first letter of a genus always appears in a capital leter.

Biological source also includes the family and the part of the drug used. For example, biological source of senna is dried leaflets of *Cassia angustifolia* Delite, belonging to family Leguminosae.

2. **Geographical Source**

   It includes the area of cultivation, collection and route of transport of a drug.

3. **Morphological Characters**

   In organized drugs, the length, breadth, thickness, surface, colour, shape, etc. are taken in morphological characters, whereas organoleptic properties (colour, odour, taste and surface) should be mentioned in unorganized drugs.

4. **Microscopical Characters**

   Microscopical characters are an important aspect of pharmacognosy as it helps in establishing the correct identity of a drug. All the detailed microscopical characters of a drug are described in this section.

5. **Chemical Constituents**

   The most important aspect which determines thetherapeutic value of a drug to which it is used is generally described in this scetion. It includes the nature, type and structure of chemical constituents present in the drug.

6. **Biological uses**

   It includes the ethnomedicinal, pharmaceutical, pharmacological and other medicinal use of the drug or the disease in which it is effective.

7. **Substituents**

   The drug which is used during non-availability of original drug is known as substituent. It has the same type of physiological active constituents; however, the quantity of the active principles may be different.

8. **Adulterants**

   With the knowledge of the diagnostic characters of drugs, the adulterants can be detected. One should have the critical knowledge of substances known to be potential adulterants. Some times the adulterants are completely devoid of physiologically active constituents, which lead in the deterioration of the quality of the drug. For example, mixing of buffalo milk with goat milk is substitution, whereas mixing of water in the milk is adulteration. In the first case, goat milk is substitute and in the second case water is adulterant.

9. **Chemical Tests**

The knowledge of chemical tests becomes more important in case of unorganized drugs whose morphology is not well defined.

3.6 Systematic position and medicinal importance of some herbs

3.6.1 Tulsi *(Ocimum tenuiflorum* L.)

Ocimum tenuiflorum (syn: *Ocimum sanctum*), commonly known as holy basil, *tulasi* (sometimes spelled *thulasi*) or *tulsi*, is an aromatic perennial plant in the Lamiaceae family. It is known as *"The Mother Medicine of Nature"* and *"The Queen of Herbs"* and is revered as an *"Alxir of Life"*. It is native to the Indian subcontinent and widespread as a cultivated plant throughout the Southeast Asian tropics.

Tulsi is cultivated for religious and traditional medicine purposes. It is widely used as herbal tea **(Fig. 11)**, commonly used in Ayurveda, and has a place within the Vaishnava tradition of Hinduism, in which devotees perform worship.

Fig. 11: ***Ocimum tenuiflorum*** **tea**

Systematic Position

Kingdom	:	Plantae
Division	:	Magnoliophyta
Class	:	Magnoliopsida
Order	:	Lamiales
Family	:	Lamiaceae
Genus	:	*Ocimum*
Species	:	*tenuiflorum*

Synonyms

Ocimum anisodorum F.Muell.

Ocimum caryophyllinum F.Muell.

Ocimum monachorum L.

Ocimum sanctum L.

Vernacular Names

Language	Vernacular Names
Sanskrit	Tulasi, Tulssi, Surasa, Krishnamul, Vishnu-Priya
Tamil	Thulasi
Hindi	Kala-tulasi
English	Holy basil
Unani	Tulsi
Bengali	Krishna tulasai

Habitat and Distribution

Tulsi is naturally found wild in the tropical and subtropical regions of the world. Basil thrives in warm and temperate climates. *Ocimum tenuiflorum* is native to India, Iran and now cultivated in Egypt, France, Hungary, Italy, Morocco and USA. In India it is indigenous to the lower hills of Punjab and Himachal Pradesh and is cultivated throughout India.

Botanical Description

Holy basil is an erect, herbaceous, branched, soft, 30–60 cm (12–24 inch) tall with tap root and hairy stems. Leaves are green to purple, simple, petiolate, with an ovate, up to 5 cm (2.0 inch) long blade which usually has a slightly toothed margin. They are strongly scented and have a decussate phyllotaxy. The purplish flowers are arranged in close whorls on elongate racemes. The two main morphotypes cultivated in India and Nepal are green-leaved (Sri or Lakshmi *tulasi*) and purple-leaved (Krishna *tulasi*). Fruits are sub-globose or broadly ellipsoid, slightly compressed, nearly smooth, pale brown or reddish with small black markings. Seeds are yellow to reddish in colour and flat.

Parts Used: Leaves, flowers, seeds, stem and dried roots

Uses of *Ocimum tenuiflorum L.* (Tulsi) in curing various ailments

The tulsi or holy basil is an important symbol in the Hindu religious tradition and is worshipped in the morning and evening by Hindus at large. The holy basil is a natural herbal remedy for the treatment of common human ailments **(Fig. 12)**. Some of the major health conditions in which the holy basil is routinely practiced are discussed under:

1. **Healing Power**

   The leaves are a nerve tonic and boosts memory. They promote the removal of the catarrhal matter and phlegm from the bronchial tube. The leaves strengthen the stomach and induce copious perspiration.

2. **Fever & Common Cold**

The leaves of basil are specific for many fevers. During the rainy season, when malaria and dengue fever are widely prevalent, tender leaves are boiled and made into tea, and given orally to cure these conditions. In case of acute fevers, a decoction of the leaves of tulsi are boiled with powdered cardamom in half a liter of water and mixed with sugar and milk and given to decrease the body temperature. In children, juice of tulsi leaves is highly recommended in bringing down the temperature during fever.

3. **Coughs**

Tulsi is an important constituent of many Ayurvedic cough syrups and expectorants. It helps to mobilize mucus in bronchitis and asthma. Chewing tulsi leaves relieves cold and flu.

4. **Sore Throat**

Water boiled with basil leaves can be taken as drink in case of sore throat. This water can also be used as a gargle.

5. **Respiratory Disorders**

The herb is useful in the treatment of respiratory system disorder. A decoction of the leaves, with honey and ginger is an effective remedy for bronchitis, asthma, influenza, cough and cold. A decoction of the leaves, cloves and common salt also gives immediate relief in case of influenza.

6. **Kidney Stones**

Basil has strengthening effect on the kidney. In case of renal stone the juice of basil leaves and honey, if taken regularly for 6 months will expel them via the urinary tract.

7. **Heart Disorders**

Basil has a beneficial effect in cardiac disease and the weakness resulting from them and reduces the level of blood cholesterol.

8. **Pediatric disorders**

Common pediatric problems like cough cold, fever, diarrhea and vomiting respond favorably to the juice of basil leaves. If pustules of chicken pox delay their appearance, basil leaves taken with saffron will hasten them.

9. **Stress**

Basil leaves are regarded as an 'adaptogen' or anti-stress agent. Recent studies have shown that the leaves afford significant protection against stress. Even healthy persons can chew 12 leaves of basil, twice a day, to prevent stress. It also purifies blood.

10. **Mouth Ulcers and Infections**

A few leaves of Basil chewed will cure and are quiet effective for the ulcer and infections in the mouth.

11. **Insect Bites**

The herb is a prophylactic or preventive and curative for insect stings or bites. A teaspoonful of the juice of the leaves is taken and is repeated after a few hours. Fresh juice must also be applied to the affected parts. A paste of fresh roots is also effective in case of bites of insects and leeches.

12. Skin Disorders

Applied topically, basil juice is beneficial in the treatment of ringworm and other skin diseases. It has also been tried successfully by some naturopaths in the treatment of leucoderma.

13. Teeth Disorder

The herb is useful in teeth disorders. Its leaves, dried in the sun and powdered, can be used for brushing teeth. It can also be mixed with mustered oil to make a paste and used as toothpaste. This is very good for maintaining dental health, counteracting bad breath and for massaging the gums. It is also useful in pyorrhea and other teeth disorders.

14. Headaches

Basil makes a good medicine for headache. A decoction of the leaves can be given for this disorder. Pounded leaves mixed with sandalwood paste can also be applied on the forehead for getting relief from heat, headache, and for providing coolness in general.

15. Eye Disorders

Basil juice is an effective remedy for sore eyes and night-blindness, which is generally caused by deficiency of vitamin A. Two drops of black basil juice are put into the eyes daily at bedtime.

16. Boosts Immunity

Holy Basil is so good for boosting up the immune system that it is hard to describe it in words. It protects against nearly all infections from viruses, bacteria, fungi, and protozoa. Recent studies show that it is also helpful in inhibiting the growth of HIV and carcinogenic cells.

17. Prevents Lung Disorders

The compounds like vitamin C, camphene, eugenol, and cineole present in the essential oils of tulsi are found to be effective in healing the damages caused to the lungs due to smoking, tuberculosis, and prevents lung cancer. It also helps to cure tuberculosis due to its antibiotic properties.

18. Prevents Premature Ageing

In the traditional Indian medicine system called Ayurveda, tulsi is considered as a tonic to retain youthful vigor and avoid premature ageing.

Tulsi Tea

Tea is made from the leaves of the plant and and has a wide range of health benefits (Fig. 12), making it a popular beverage around the world. An active ingredient, Eugenol present in its leaves, delivers the major benefits. Tulsi tea is considered healthy and has been an important component of Ayurveda for thousands of years.

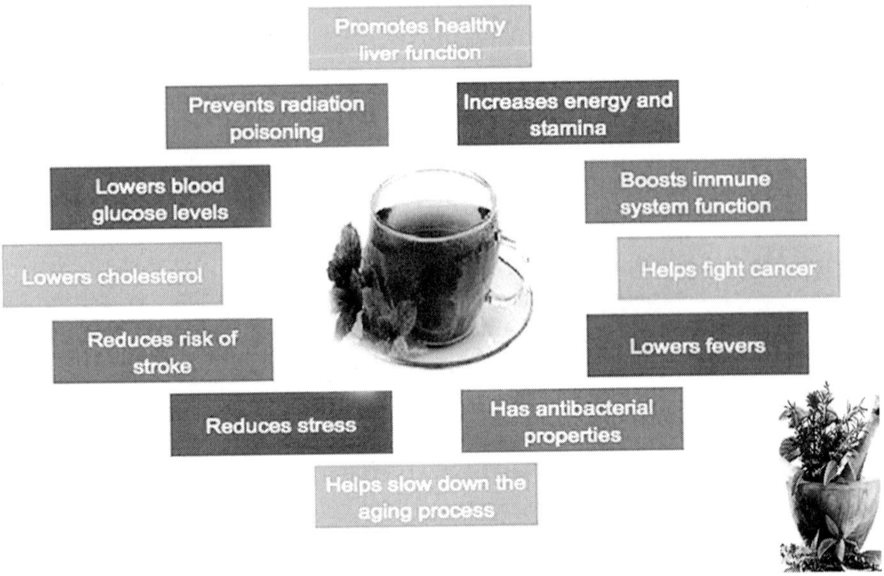

Fig. 12: Major health benefits of Tulsi tea

3.6.2 Indian Gooseberry *(Phyllanthus emblica* L.)

Phyllanthus emblica, also known as emblic, emblic myrobalan, Indian gooseberry or Amla from Sanskrit Amalaki is a deciduous tree of the family Phyllanthaceae. It is known for its edible fruit of the same name.

Fig. 13: *Phyllanthus emblica* with fruits

Systematic Position

Kingdom	:	Plantae
Division	:	Magnoliophyta
Class	:	Magnoliopsida
Order	:	Malphigiales
Family	:	Phyllanthaceae
Genus	:	*Phyllanthus*
Species	:	*emblica*

Synonyms

Emblica arborea Raf.

Emblica officinalis Gaertn.

Phyllanthus taxifolius D.Don

Phyllanthus mairei H. Lev.

Phyllanthus mimosifolius Salisb.

Cicca emblica (L.) Kurz

Diasperus emblica (L.) Kuntze

Vernacular Names

Country	Vernacular Names
England	Emblic, Emblic myrobalan, Myrabolan
UK	Emblique officinale
China	Dianganlan, ganlanshu
Germany	Amalanbaum
India	Amla, Amlika, Usiribi
Indonesia	Kimalaka
Nepal	Amala
Phillippines	Nelli
Thialand	Ma-Khaam-pom
Malaysia	Malaka

Geographic distribution

Phyllanthus emblica is widely distributed in most tropical and subtropical countries including China, Indonesia, and on the Malay Peninsula. *P. emblica* is native to tropical southeastern Asia, particularly in central and southern India, Nepal, Pakistan, Bangladesh, Bhutan, Sri-lanka, Malaya, Myanmar and the Mascarene Islands. It is abundant in deciduous forests of Madhya Pradesh. Originally, it was cultivated in Madascar.

Botanical description

The tree is small to medium in size, reaching 1–8 m (3 ft 3 in–26 ft 3 in) in height. The branchlets are not glabrous or finely pubescent, 10–20 cm (3.9–7.9 in) long, usually deciduous; the leaves are simple, sub sessile and closely set along branchlets, light green, resembling pinnate leaves. The flowers are unisexual and greenish-yellow. The fruit is nearly spherical, light greenish-yellow, quite smooth and hard, with six vertical stripes or furrows **(Fig. 13)**. The fruit diameter ranges between 1.8 and 2.5 cm.

Parts used: Leaves, roots, seeds, fruits, bark and flowers.

Medicinal uses of *Phyllanthus emblica* (Indian Gooseberry)

Fruits of *P. emblica* have been used for thousands of years in the traditional Indian medicine for the treatment of several diseases. All parts of the plant are used for medicinal purposes, especially the fruit. In traditional medicinal system, dried and fresh fruits of amla are frequently used as source of vitamin C, liver, digestive tonic, immunity booster etc. **(Fig. 14)**.

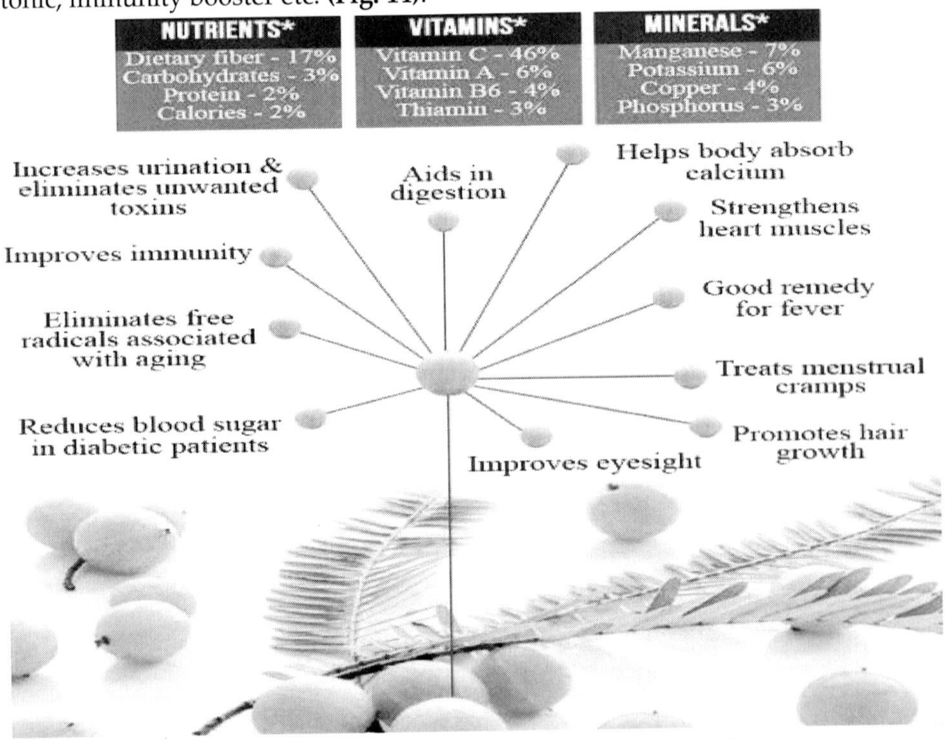

Fig. 14: Nutritional value and health benefits of Amla fruits

1. **Hair fall**

   Amla is good to prevent hair loss, hair fall and plays a vital role in the overall health of hair. Regular intake of Amla juice makes hair darker and thicker and also prevents greying. Amla prevents baldness due to the presence of carotene and iron. The mixture of juices of amla and lemon when applied over the scalp and left it for 20-30 minutes, helps to strengthen the hair follicles and shaft, enhances hair lustre, color and pigmentation.

2. **Skin problems**

   Drinking of juice is beneficial to make skin glow, by removing the dead cells. The application of amla juice over the face helps to fight wrinkles, pimples, acne etc.

3. **Eye Disorders**

   Amla prevents many eye related problems such as cataract, near-sightedness, vision improvement etc. The presence of vitamin A and carotene overcomes many problems related with eyes.

4. **Cough and Cold**

   The mixture of fresh amla juice and honey, if taken daily, helps to eradicate the cough and cold.

5. **Mouth Ulcers**

   Amla juice (2 tbsp) is added in half glass of water and gargled to cure mouth ulcers.

6. **Diabetes**

   The regular consumption of amla juice helps to lower the levels of sugar in blood thus controls diabetes. Amla juice stimulates the islet of langerhans, which further regulate the optimum secretion of insulin.

7. **Indigestion**

   It is good for digestion and assimilation of foods. It has adequate amount of fibre and roughage that makes one relieve from constipation and indigestion. It ensures to flush out toxins from the body, balances and strengthens the digestion process.

8. **Asthma**

   Taking amla juice and honey on regular basis helps to control asthma, bronchitis, allergic asthma, respiratory congestion, chronic cough and cold.

9. **Weight loss**

   Amla is good to burn fat, as it has the high level of amino acid that helps to shed unnecessary fats from the body.

10. **Cardiac disorders**

    Regular intake of amla juice maintains heart healthy and fit. It lowers the level of cholestrol and strengthens the cardiac muscles thus smoothens the pumping of blood and oxygen to the entire body. Presence of chromium prevents plaque formation in the blood vessels, thus prevents strokes and other cardiac problems.

3.6.3 Ginger (*Zingiber officinale Rosc.*)

Ginger (*Zingiber officinale* Rosc.) is a flowering plant belongs to the family *Zingiberaceae*, which is composed of 50 genera and around 1500 species of perennial tropical herbs. The English botanist William Roscoe gave the plant the name

'*Zingiber officinale*' in 1807. The name '*Zingiber*' is via the Greek word '**zingiberis**' derived from the Sanskrit word '**shringavera**', which means 'shaped like a deer's antlers'; '*officinale*' indicates medicinal properties of the plant

Systematic Position

Kingdom	:	Plantae
Division	:	Magnoliophyta
Class	:	Lilopsida
Order	:	Zingiberales
Family	:	Zingiberaceae
Genus	:	*Zingiber*
Species	:	*officinale*

Synonyms

Zingiber albiflorum R.M.Sm.

Zingiber capitatum Roxb.

Zingiber officinale var. *cholmondeleyi* F.M. bailey

Zingiber officinale var. *rubrum* theilade

Zingiber officinale var. *macrorhizonum* Makino

Zingiber officinale f. *macrorhizonum* (Makino) M.Hiroe

Vernacular names

Language	**Vernacular Name**
Hindi	Adrak
Manipuri	Shing
Marathi	Alha, Aale
Tamil	Ingee, Inji
Malayalam	Inchi, Enchi
Telugu	Allam, Allamu, Shonti
Bengali	Ada
Oriya	Ada
Urdu	Adrak, Adi
Sanskrit	Adraka
Nepali	Aduwa

Geographical Distribution

Ginger is indigenous to Indo-Malayan region and has been in cultivation in India since prehistoric times. It is now cultivated over a greater part of the tropical and temperate regions. The chief ginger growing countries of the world are India, China, Australia, Dutch East Indies, West Indies, Mexico, Jamaica, North Africa and West Africa.

Botanical Description

Ginger is an aromatic herb with an underground rhizome and an erect stem, up to 75 cm in height **(Fig. 15)**. Leaves of ginger are simple, alternate, linear-lanceolate, sheathing at the base, sessile, acuminate at apex, glabrosus, up to 15 cm long. Inflorescence is a spike on a distinct scape; flowers densely arranged, bisexual, irregular, each subtended by a persistent scarious bract.

Fig. 15: Different parts of Ginger plant

Fruit is an oblong capsule, many seeded, globose, with a small embryo and copious endosperm.

Parts Used: Rhizome

Ginger rhizome is widely consumed as a spice in foods and beverages because of its characteristic pungent flavour.

Medicinal uses of *Zingiber officinale Rosc.* (Ginger)

Ginger is a popular ingredient in cooking, and especially in Asian and Indian cuisine. The underground stem or rhizome of the ginger plant is commonly known as the ginger root and is well known for its medicinal properties. Ginger has been used as a natural remedy for thousands of years across the globe. Ancient Chinese, Roman, Greek, Arabic and Sanskrit texts have documented the use of ginger root to improve health and well-being. Ginger is associated with an extensive health benefits.

1. **Indigestion**

   The phenolic compounds in ginger are known to help relieve gastrointestinal (GI) irritation, stimulate saliva and bile production, and suppress gastric contractions as food and fluids move through the GI tract.

2. **Nausea**

   Chewing raw ginger or drinking ginger tea is a common home remedy for nausea during cancer treatment. Taking ginger for motion sickness reduces the feelings of nausea.

3. **Cold and flu**

   During cold weather, drinking ginger tea is good way to keep warm. It is diaphoretic, which means that it promotes sweating, working to warm the body from within.

4. **Pain**

   Daily ginger supplementation reduced exercise-induced muscle pain. Ginger has also been found to reduce the symptoms of dysmenorrhea, the severe pain that some women experience during a menstrual cycle.

5. **Inflammation**

   Ginger has been used for centuries to reduce inflammation and treat inflammatory conditions. Ginger has also been found to be "modestly efficacious and reasonably safe" for treating inflammation associated with osteoarthritis.

6. **Cardiovascular diseases**

   The use of ginger reduces cholesterol and lowers the risk of blood clotting, and helps to maintain healthy blood sugar levels.

7. **Prevent Cancer**

   Ginger extract has been used as an alternative treatment for several forms of cancers. The anticancer properties of ginger are attributed to the presence of 6-gingerol compound.

8. **Improves brain function**

   Studies suggest that ginger extract has a potential to protect damage to the brain by inhibiting the inflammatory responses that occur in the brain. It can also improve age-related decline in brain function in elderly ones.

9. **Fight infections**

   Gingerol, a bioactive substance present in the raw ginger can lower the risk of several bacterial infections. It is highly effective against the oral bacteria linked to diseases of gums like gingivitis, periodontitis etc. Fresh ginger is also effective against the common viral respiratory infections.

10. **Protects against stomach ulcers**

    Stomach ulcers are the painful sores in the lining of the stomach. Several studies have shown that ginger is helpful in preventing the formation of stomach ulcers. Ginger extracts cures the stomach ulcers by reducing the levels of inflammatory proteins and blocking the activity of enzymes related to ulcer development.

3.6.4 Ashoka Tree (*Saraca asoca* Roxb. Willd.)

Saraca asoca (The **Ashoka tree**; lit., "**sorrow-less**") is a plant belonging to the Caesalpinioideae subfamily of the legume family. Ashoka is a sanskrit word meaning *"without grief"*. It is one of the most legendary and sacred trees of India. It is an important tree in the cultural traditions of the Indian subcontinent and adjacent areas. It is sometimes incorrectly known as *Saraca indica.* The plant grows faster in tropical and sub-tropical situation. All parts of the tree such as bark, leaves, flowers and seeds are used medicinally. Due to its medicinal value, it is known as a universal plant.

Systematic Position

Kingdom	:	Plantae
Division	:	Magnoliophyta
Class	:	Magnoliopsida
Order	:	Fabales
Family	:	Fabaceae
Genus	:	*Saraca*
Species	:	*asoca*

Synonyms

Jonesia asoca Roxb.

Jonesia confusa Hassk.

Saraca confusa (Hassk.) Backer

Vernacular names

Language	Vernacular name
Hindi	Ashok, Sita Ashok
Gujarati	Ashopalava
Kannada	Achenge
Malayalam	Hemapushpam
Marathi	Jasundi
Tamil	Asogam
Telugu	Asokamu
Nepali	Ashok, Ashau

Habitat and distribution

Ashoka (*Saraca asoca*) is a rain forest tree. It is native of Asia and South America. This tree is originally distributed in the central areas of Deccan Plateau. It is also found in the Western Ghats of the Indian subcontinent, in the center and the

Eastern Himalayas and in the hills of Khasi, Garo and Lushaj. Ashoka is indigenous to India, Burma and Malaya. In Indian, it is distributed in West Bengal, Kerala, Maharashtra, Andhra Pradesh and Meghalaya.

Botanical Description

Ashoka (*Saraca asoca*) is a small evergreen tree. The leaves are alternate, paripinnate compound, stipulate, 7-30cm long, large and spreading horizontally. The flowers are orange coloured with a beautiful aroma. 7-8 stamens are found in flower **(Fig. 16)**. Fruits are smooth, leathery and flat pods including 6-8 seeds inside. Bark is externally dark green to greenish grey in colour. It is smooth with transversely ridged and circular lenticels.

Fig. 16: Photograps showing Ashoka tree, flowers and fruits

Parts Used: Stem, Bark, Seeds, Flowers

Medicinal uses of *Saraca asoca* (Ashoka Tree)

The meaning of Ashok in Sanskrit is without grief/ sorrow less. Each part of the tree is of proven medicinal value and cures many disorders, hence it is called Ashoka.

The tree has many medicinal qualities. Needless to say, dried stem, bark, seeds and flowers from the tree, are widely used in India for treating a variety of health conditions.

Some of the amazing benefits of Ashoka tree are discussed below:

1. **Gynecological Problems**

   Ashoka tree is popularly used in the treatment of gynecological problems and menstrual disorders in women. The tree acts as a tonic for uterine muscles and endometrium, controlling spasms and abdominal pain.

2. **Skin Complexions**

   The Ashoka tree naturally improves the skin's complexion. It helps in blood purification and prevention of further skin allergies. As it removes toxins, it serves to clear the complexion. Moreover, extracts from the tree can also be used to soothe burns and skin irritations.

3. **Pain**

   Ashoka tree has pain relieving properties and can be used as an analgesic safely. The paste of the bark is used to treat joint pains.

4. **Internal Bleeding**

   The tree is effectively used to treat internal bleeding and hemorrhages. The flower extract of this plant is used to treat hemorrhagic dysentery.

5. **Diabetes**

   Dried flowers are used to cure diabetes. It adds ease for indigestion. Consume tonics or decoctions regularly to benefit from the tree's offerings.

6. **Piles**

   Ashoka tree extracts are used in the treatment of internal piles and bleeding due to piles. The herbal supplements, prepared from the bark, are made into decoctions and consumed.

7. **Infections**

   Ashoka's bark effectively cures fungal and bacterial infections. The consumption of bark purifies the body and eliminates infection.

8. **Stomach swelling and inflammation**

   Leaves and bark is consumed to clear the stomach from worm infestations. It not only relieves pain, but also reduces bloating.

9. **Diarrhea**

   Leaves, flowers and bark are used to treat diarrhea without any side effects. It also purifies the blood in the process.

10. **Kidney Stones**

    Ashoka tree seed powder is a natural method to control kidney stones. Seed powder is boiled in water and taken not only to dissolve but to ease the removal of kidney stones.

3.6.5 Fenugreek (*Trigonella foenum-graecum* L.)

Trigonella foenum-graecum, commonly known as Fenugreek, is an ancient spice and popular in Arabian regions and India. It is one of the plants used as an herb (leaves) and as a spice (seeds). The dried leaves have a bitter taste and a strong characteristic smell, used as a flavoring agent in many dishes, curry in particular. Fresh tender pods, leaves and shoots are eaten as vegetables, while seeds

Fig. 17: *Trigonella foenum-graecum*

are used for flavoring variety of foods. The name fenugreek derives from "*Greek hay*", illustrating its classical use as a fodder. Bitterness in the seeds is mainly due to the oil, steroidal saponins and alkaloids.

Systematic Position

Kingdom	:	Plantae
Sub-kingdom	:	Tracheobionta
Super-division	:	Spermatophyta
Division	:	Magnoliophyta
Class	:	Magnoliopsida
Subclass	:	Rosidae
Order	:	Fabales
Family	:	Fabaceae
Genus	:	*Trigonella*
Species	:	*foenum-graecum*

Vernacular names

Language		**Vernacular name**
English	:	Fenugreek, Greek hay
Hindi	:	Methi
Sanskrit	:	Methika
Tamil	:	Vendhayam
Telugu	:	Menthulu

Botanical description

It is an erect hairy annual herb of the bean family, reaching 30-60 cm. The plant grows to a height of about 3 feet, has three part leaves **(Fig. 17)**, the long slender stems bear tripartite, toothed, grey green obovate laves, 20-25 mm long. *Trigonella foenum-graecum* has long stalked leaves upto 5 cm long stipules triangular, lanceolate, leaflets about 2.5cm long, obovate to obanceolate. The root is a mass of fingery structures. The sessile axillary flowers are white or pale yellow. The thin, sword shaped pods are 10-15 cms (4-6 in), with a curved beak like tip, each carrying 10-20 seeds. The plant radiates a spicy odour which persists on the hands after touching. Seeds are small (5mm long), hard and brownish yellow (The colour may vary).

Medicinal uses of Fenugreek *(Trigonella foenum-graecum* L.)

Parts used: Leaves and Seeds

Fenugreek has been used to treat a long list of diseases and conditions in alternative medicine. Preliminary research suggests that fenugreek can help with the following:

1. **To enhance milk production in new mother**

   Fenugreek is widely used as a galactagogue or a milk flow enhancing agent in new mothers. Nursing mothers are fed with fenugreek pills or tea to enhance lactation.

2. **To maintain blood sugar levels**

   Fenugreek seeds are commonly used as a supplement to control blood glucose, especially to prevent or treat diabetes.

3. **To boost libido**

   One of fenugreek's ancient use is to enhance libido. Recent studies have suggested that fenugreek may increase libido in both men and women.

4. **To soothe skin inflammation or injury**

   Fenugreek powder combined with other soothing herbs as a poultice is applied directly to the skin to treat swelling (inflammation), muscle pain and swelling of lymph nodes (lymphadenitis), pain in the toes (gout), wounds, leg ulcers and eczema.

5. **Painful menstrual periods**

   Fenugreek seed powder is taken thrice daily to reduce pain in women during menstrual periods. Fenugreek seed powder is a safe and effective alternative to pain killers during pain associated with periods.

6. **Aids digestion**

   It purifies blood and helps in flushing out the harmful toxins, thereby making the digestive system refreshed and clean.

7. **Prevents hair loss**

   The fenugreek seeds being high source of proteins are useful in the prevention of hair fall, baldness and thinning of hair.

8. **Cures skin problems**

   Fenugreek seeds prove to be an excellent beauty product. They help to prevent wrinkles, blackheads, pimples, dryness and rashes.

9. **Weight loss**

   Fenugreek seeds are found to be helpful in loosing excess weight, due to its high fibre content. Seeds are soaked in water for overnight and chewed in the early morning to reduce excess weight.

10. **Nutritional value**

    Traditionally, fenugreek seeds in the form of seeds, sprouted, powder, sauce or as paste are in use in a wide range of culinary recipes due to its high nutritional ingredients. Seeds are a rich source of minerals, vitamins and other nutrients **(Table 7).**

Table 7: Nutritional value of Fenugreek per 100 g
(Source: USDA National Nutrient data base)

Principle	Nutrient value
Carbohydrates	58.35g
Proteins	23g
Lipids	6.41g
Dietary fiber	24.6g

Principle	Nutrient value
Vitamins	
Folates	57 μg
Niacin	1.640 mg
Pyridoxine	0.600 mg
Riboflavin	0.366 mg
Thiamin	0.322 mg
Vitamin C	3 mg
Electrolytes	
Sodium	67 mg
Potassium	770 mg
Minerals	
Calcium	176 mg
Copper	1.110 mg
Iron	33.53 mg
Magnesium	191 mg
Manganese	1.228 mg
Phosphorus	296 mg
Zinc	2.50 mg

3.7 Future of Pharmacognosy

Medicinal plants are of great value in the field for treatment and curing of many diseases. Over the years, scientific research has expanded our knowledge of the chemical effects and composition of the active constituents, which determine the medicinal properties of the plants. It has now been universally accepted fact that the plant drugs and remedies are so safer than that of synthetic medicines for curing the complex diseases like Cancer and AIDS. Enormous number of alkaloids, glycosides and antibiotics have been isolated, identified and used as curative agents. The modern developments in the instrumental techniques of analysis and chromatographical methodologies have added numerous complex and rare natural products to the armoury of phytomedicine like artemisin as antimalarial, taxol as anticancer, forskolin as antihypertensive, rutin as capillary permeability factor and piperine as bioavailability enhancer. Natural products have also been used as drug substitutes for the semi synthesis of many potent drugs.

In the Western world, as the people are becoming aware of the potency and side effects of synthetic drugs, there is an increasing interest in the plant based remedies with a basic approach towards the nature. The future developments of Pharmacognosy as well as herbal drug industry would be largely dependent upon the reliable methodologies for identification of marker compounds of the extracts and also upon the standardisation and quality control of these extracts. Mother earth has given vast resources of medicinal flora and fauna both terrestria and marine, and it largely depends upon the forthcoming generations of pharmacognosists and phytochemists to explore the wonder drug molecules from this unexploited wealth.

Little more needs to be said about the present day importance of medicinal plants, for it will be apparent from the foregoing that the plant themselves either in the form of crude drugs or even more important, for the medicinally active materials isolated from them, have been, are and always will be an important aid to the physician in the treatment of many diseases.

UNIT - IV

Analytical Pharmacognosy

4.1 Drug Evaluation

Herbal drugs have been widely used for the thousands of years in developing and developed countries owing to its natural origin and lesser side effects or dissatisfaction with the results of synthetic drugs. However, quality control of oriental herbal drugs is more difficult than that of western drug, due to their composite nature. Therefore, every single herbal drug needs to be quality checked to ensure that it confirms to quality requirement and delivers the properties consistently. Standardization assures that products are reliable in terms of quality, efficacy, performance and toxicity.

Defnition

The process by which toxicity, safe dosage range, quality and purity and efficacy of drug or group of drugs is determined is called drug evaluation.

Identity-identification of biological source of the drug

Quality- the quantity of the active constituents present

Purity- the extent of impurities present in the drug

Need and importance of evaluation of drugs

The evaluation of plant-derived drug is essential because:

- there may be biochemical variation in the drugs
- of deterioration due to treatment and storage and
- of substitution and adulteration, as a result of carelessness, ignorance etc.

For the quality control of a traditional medicine, the traditional methods are procured and studied, and documents and the traditional information about the identity and quality assessment are interpreted in terms of modern assessment or monograph in herbal pharmacopoeia.

4.1.1 Methods of Drug Evaluation

The plant-based drugs are identified on the basis of their morphological, histological, anatomical, embryological and chemical characters. Based on these characters, different techniques are used to evaluate and standardrize the drugs.

1. **Morphological and Organoleptic Evaluation**

   Drugs are evaluated by their morphological or macroscopic features like color, size, shape, odor and special features like texture, touch etc. It is the qualitative evaluation technique based on the study of morphological and sensory profile of whole drugs.The fractured surfaces in cinchona, quillia and cascara barks and quassia wood are important charactersistics. Aromatic odour of umbelliferous fruits and sweet taste of liquorice are the examples of this type of evaluation where odor of drug depends upon the type and quality of odourous principles (volatile oils) present. Shape of drug may be cylindrical (sarsapilla), subcylindrical (podophyllum), conical (aconite), fusiform (jalap) etc. The general appearance of the weight of a drug often indicates whether it is likely to comply with prescribed standard like furrows (alternate depression), wrinkles (fine delicate furrows), annulations (transverse rings), fissures (splits), nodules (rounded outgrowth), scars (spot left after fall of leaves, stems or roots). Size represent length, breadth, thickness, diameter etc.; color means external color which varies from white to brownish black; taste may be acidic (sour), saline (salty), saccharic (sweetish), bitter or tasteless (no taste) are the important diagnostic characters.

2. **Microscopic Evaluation**

Microscopic investigation allows more detailed examination of a drug by virtue of its magnification power. Drugs are evaluated by their known histological characters like stomatal index, palisade ratio, trichomes, vascular tissues, starch granules, calcium oxalate crystals, aleurone grains etc. It is one of the most important qualitative evaluations of organized crude drugs in entire or powdered forms with help of microscope.

Herbal drugs can also be identified microscopically by cutting the thin TS (transverse section), LS (Longitudinal section), especially in case of wood and by staining them with proper staining reagents e.g. starch and hemicelluloses are identified by blue color with iodine solution.

Quantitative aspects of microscopy includes study of stomatal number and index, palisade ratio, vein-islet number, size of starch grains, length of fibers etc. which plays an important role in the identification of drugs.

3. **Physical Methods**

Physical parameters are sometimes taken into consideration to evaluate certain drugs. These include moisture content, specific gravity, optical rotation, refractive index, melting point, viscosity and solubility in different solvents. All these physical properties are useful in the identification of constituents present in the plant based drug.

4. **Chemical Evaluation**

Most of drugs have definite chemical constituents to which their biological or pharmacological activity is attributed. The isolation, purification, identification of active constituents present in the drug is based on chemical methods of evaluation. Different chemical tests and assays are employed to determine chemical constituents present in the drug. When the active constituents (e.g. sennosides in *Senna*) or markers (e.g. alkylamidesin *Echinacea*) are known, a vast array of modern chemical analytical methods such as Ultraviolet/Visible Spectroscopy (UV/VIS), Thin Layer Chromatography (TLC), High Performance Liquid Chromatography (HPLC), Gas Chromatography (GC), Mass Spectrometry (MS), or a combination of GC and MS (GC-MS), Nuclear Magnetic Resonance (NMR), Fourier Transform-Infra Red Spectroscopy (FT-IR) etc. are employed to identify and characterize the compounds present in the drug.

Quantitative chemical analysis such as saponification value, acid value, phenolic and flavonoid, alkaloid, terpenoid, saponin contents are also covered under chemical evaluation. Some of these chemical analytical tests are useful in evaluation of resins (acid value, sulphated ash), balsams (acid value, saponification value and bester values) and volatile oils (acetyl and ester values).

The major and the most commonly used techniques under chemical evaluation of drugs are summarised below:

- Nuclear Magnetic Resonance (NMR) Spectroscopy: NMR is an important tool for illustration of molecular structure, especially the stereochemistry and configuration.
- Infra-Red (IR) Spectroscopy: It measures absorbed, transmitted or reflected radiant energy in the region of elcetromagnetic spectrum, ranging from 4000 to 400 cm^{-1}. IR spectroscopy determines the chemical compounds and identifies the active groups in the herbal drugs.
- Mass Spectroscopy: It is an analytical technique which determines the molecular weight, structure and chemical nature of compounds in the herbal formulations.
- Thin layer chromatography: TLC/HPTLC is frequently used for the analysis of herbal medicines since various pharmacopoeias such as Indian herbal pharmacopoeia, Ayurvedic pharmacopoeia; American Herbal Pharmacopoeia (AHP), Chinese drug monographs and analysis, Pharmacopoeia of the People's Republic of China etc. TLC has the advantages of many-fold possibilities of detection in analyzing herbal medicines. The advantages of using TLC/HPTLC to construct the fingerprints of herbal medicines are its simplicity, versatility, high velocity, specific sensitivity and simple sample preparation.
- High-performance liquid chromatography: HPLC is a popular method for the analysis of herbal medicines because it is easy to learn and use and is not limited by the volatility or stability of the sample compound. In general, HPLC can be used to analyze almost all the compounds in the herbal medicines. Thus, over the past decades, HPLC has received the most extensive application in the analysis of herbal medicines. Reversed-phase (RP) columns may be the most popular columns used in the analytical separation of

herbal medicines

- Gas Chromatography: Many pharmacologically active components in herbal medicines are volatile chemical compounds, thus, the analysis of volatile compounds by gas chromatography is very important in the analysis of herbal medicines. the GC of the volatile oil gives a reasonable "fingerprint" which can be used to identify the plant. The composition and relative concentration of the organic compounds in the volatile oil are characteristic of the particular plant and the presence of impurities in the volatile oil can be readily detected.
- GC-MS: With the GC–MS, not only a chromatographic fingerprint of the essential oil of the herbal medicine can be obtained but also the information related to its most qualitative and relative quantitative composition is determined. In the herbal medicine analysis, GC coupled mass spectroscopy, is extremely useful in elucidating the relationship between chemical constituents in herbal medicine and its pharmacology. Thus, GC–MS is one of the most preferable tools for the analysis of the volatile chemical compounds in herbal medicines.
- Flourescence Analysis: Natural products produce different flourescence in visibile, UV light. The drug belladona shows flourescence in visible range, while cinchona produces flourescence in UV.

5. **Biological Evaluation**

Plant based drugs are the repository of active agents, which have specific pharmacological targets. The efficacy o f the crude drugs or its formulations are evaluated against microbes like bacteria, fungi, viruses, protozoans and other animals by using different bioassays. For biological evaluation of drugs, experiments are carried out on both intact and isolated organs of living animals. With the help of bioassays (testing the drugs on living animals), strength of drug is being evaluated. Biologically, drugs are evaluated for antimicrobial, antioxidant, antifertlity, antiulcer, anticancer, antidiabetic, anti-inflammatory, analgesic, wound healing, hypoglycemic, antidepressant, diuretic, hypolipidemic, hepatoprotective, anti-diarrheal, antiepileptic, insecticidal, larvicidal, spasmolytic activities.

4.2 Phytochemical screening tests

Phytocompounds represent a diverse group of chemical classes, including alkaloids, flavonoids, anthraquinones, coumarins, gums, glycosides (cardiac glycosides and cyanogenic glycosides), phenols, saponins, tannins, polysaccharides, terpenes and terpenoids present in the plants. Phytochemical estimation of medicinal plants is an essential step in the field of plant research. Analysis of phytochemicals largely depends on the application of bioassay. A variety of standard methods are currently available for the screening of crude drugs and isolated phytocompounds for their selective chemical constituents.

Phytochemical screening tests determine the presence of a particular chemical group in the crude drug.

4.2.1 Tests for Alkaloids

Mayer's test

About 0.5-1ml of plant sample is taken in a test tube. Few drops of Mayer's reagent are added. It is shaken and allowed to stand for some time. Appearance of cream colour precipitate indicates that alkaloids are present.

Hager's test

2 ml alcoholic plant extract is taken in a test tube and a few drops of Hager's reagent are added. Formation of yellow coloured precipitate indicates the presence of alkaloids in the sample.

Dragendorff's test

To 2 ml of alcoholic plant extract 5 ml of distilled water is added in a test tube. To this mixture few drops of hydrochloric acid (2 M) are added and shaken well. Then, 1 ml Dragendorff's reagent is added to the mixture, formation of orange or orange-red precipitate indicates the presence of alkaloids in the sample.

Wagner's test

About 2-3 ml alcoholic plant extract is taken in a test tube and a few drops of hydrochloric acid are added. Then few drops of Wagner's reagent are added, shaken and allowed to stand for some time. Appearance of yellow or brown coloured precipitate indicates that alkaloids are present.

4.2.2 Tests for Flavonoids

Alkaline reagent test

Few drops of NaOH solution (10%) are added to 0.5-1ml of test solution. Formation of an intense yellow colour, which turns colourless on addition of two drops of dilute acid, indicates presence of flavonoids.

Shinoda test

2 ml of ethanolic plant extract is taken in a test tube and 10 ml of dilute hydrochloric acid is added. A small piece of magnesium is added to the mixture, formation of red or pinkish colour in the solution indicates that flavonoids are present in the plant sample.

4.2.3 Tests for Steroids and Triterpenoids

Salkowski test

About 0.5-1ml of sample is allowed to react with $CHCl_3$. Few drops of concentrated H_2SO_4 are added and shaken vigorously. After some time, if red colour appears at the lower layer, it confirms that steroids are present, while the formation of yellow lower layer confirms the presence of triterpenoids in the sample.

Liebermann-Burchard's test

2 mg of dry ethanolic plant extract is taken in a test tube and treated with 5 ml acetic anhydride. The resulting mixture is then boiled and then allowed to cool

at room temperature. To this mixture, 1 ml of concentrated H_2SO_4 is added drop by drop along the sides of the test tube. Formation of green colour indicates the presence of steroids in the plant sample, while appearance of pink colour indicates that triterpenoids are present.

4.2.4 Tests for Tannins

Ferric chloride test

To 0.5ml of plant sample, 2 to 3 drops of $FeCl_3$ are added. The mixture is allowed to react and if bluish-green colour appears, it confirms that tannins are present in the sample. The formation of green colour indicates the presence of gallotannins while light brown colour indicates presence of pseudotannins.

4.2.5 Tests for Saponins

Foam test

2ml of test solution is vigorously shaken with equal amount of distilled water in a test tube. If foam appears and persists for nearly 10 minutes, it shows that the sample is having the saponins.

Froth test

To 5 ml of ethanolic plant extract, a drop of sodium bicarbonate is added in a test tube. The mixture is vigorously shaken and left undisturbed for 3 minutes. Development of froth in the upper part of the test tube indicates the presence of saponins in the sample.

4.2.6 Tests for Carbohydrates

Benedict's test

About 0.5 to 1ml of plant sample is reacted with few drops of Benedict's reagent. Upon boiling, reddish-brown ppt. forms, which shows that reducing sugars are present in the sample.

Molisch's test

About 2 ml test solution (1 gm plant powder and 5 ml distilled water) is taken in a test tube and treated with 2 -3 drops of ethanolic solution of α-naphthol. To this 1 ml concentrated H_2SO_4 is added along the wall of the test tube. If purple or violet coloured ring appears, indicates the presence of carbohydrates in the plant sample.

Fehling's test

In a test tube, 2 ml ethanolic plant extract and 10 ml distilled water is taken and shaken well. The mixture is filtered and the filtrate is taken in the test tube and 1 ml each Fehling's solution A and Fehling's solution B is added. The mixture is boiled for about 5 minutes and allowed to stand for some time. The formation of red or brick-red precipitate indicates the presence of carbohydrates in the plant sample.

Anthrone test

2 mg of ethanolic plant extract is added to 10 ml distilled water and shaken well. The solution is filtered and filtrate is taken in a test tube. To the filtrate 2 ml

anthrone reagent is added, bluish-green colour develops, indicates the presence of carbohydrates in the sample.

4.2.7 Tests for Proteins and Amino acids

Ninhydrin test

0.2% Ninhydrin solution is reacted with 1-2 ml of plant sample. The mixture is boiled and allowed to cool. Occurrence of violet colour confirms that proteins are present in the sample.

Millon's test

About 1 gm plant powder is added in 5 ml distilled water and mixed well. To this mixture 1 ml Millon's reagent (15% mercuric sulphate solution + 15% H_2SO_4) is added and heated for 15 minutes and allowed to cool at room temperature. After that 1 % sodium nitrate solution is added to the mixture, formation of white precipitate indicates the presence of amino acids in the plant sample.

4.3 Drug adulteration

Adulteration may be defined as mixing or substituting the original drug material with other spurious, inferior, defective, spoiled or other parts of same or different plant. The adulteration of herbal drugs is the burning issue in herbal industry which has caused a major effect in the commercial utilization of natural products.

Drug adulteration methods

Adulteration may generally take place by two ways: 1) Direct or intentional adulteration and 2) Indirect or unintentional adulteration

1. **Direct or intentional adulteration**

   Due to morphological resemblance to the authentic herb, many different inferior commercial varieties are added intentionally as adulterants **(Table 8)**, which may not have any therapeutic potential. The practice of intentional adulteration is mainly encouraged by traders who are unwilling to pay extravagant price for herbs of superior quality, and hence are supposed to purchase only the cheaper products. This encourages producers and traders to sell herbs of inferior quality.

Table 8: Adulterants found in some of the major traded medicinal products

S. No.	Commodity	Adulterants	
		Non-biological	Biological
01	Ginger powder	Lime	Capsicum; turmeric; exhausted ginger fortified with flavours; Japanese ginger
02	Black pepper powder	Dye	Powdered papaya seeds; wild *Piper* berries; *Lantana camara*; *Embelia ribes*; *Mirabilis jalapa* seeds; *Schinus* molle berries; exhausted black pepper and light berries

S. No.	Commodity	Adulterants	
		Non-biological	Biological
03	Chilli powder	Dye-coal tar red, sudan red, para red; vanilyl-n-nonamide; Mineral oil; talc powder; brick powder; salt powder	Powder fruits of 'Choti ber' (*Ziziphus nummularia*); red beet pulp; almond shell dust; extra amounts of bleached pericarp, seeds, calyx and peduncle of chilli; tomato wastes
04	Turmeric powder	Dye Mentanil Yellow, Orange II lead chromate; chalk powder; brick powder	Wild *Curcuma* spp, *C zeodaria* Rose or 'yellow shotti' syn. *C. malabrica;* saw dust
05	Cinnamon powder	Eugenol, cylon oil, yellow brown dye	Aromatized and powdered beechnut husk; hazel nut; almond shell dust
06	Cardamom fruits	Small pebbles	Orange seeds; unroasted coffee seeds
07	Cardamom seeds	-	Seeds of *Amomum aromaticum, A. subalatum* and *A. cardamomum*
08	Nutmeg	-	Wild species of Macassar (*Myristica argentea*), Bombay nutmeg (*M. malabarica*) and *M. otoba*
09	Clove	Magnesium salt, sand, earth	Exhausted cloves stem and fruits of clove
10	Mustard seeds	-	Argemone seeds (*Argemone mexicana*); rape seed; ragi
11	Mustard seed powder	-	Added starch in turmeric
12	Cassia bark	-	Bark of *Cinnamomum japonicum, C. mairei, C. Burmanii*
13	Allspice powder	-	Powdered clove stem; berries of *Myrtus tobasco* and *Lindera benzoin*
14	Aniseed	Fine earth metals	Hemlock fruit; parsley; dill fruit
15	Nigella seeds	-	Onion seeds
16	Caraway	-	Cumin; *Carum bulbocastanum*
17	Fennel	-	Exhausted or partially exhausted fennel fruits; stem tissue and stalks of fennel; umbelliferous seeds
18	Poppy seed	-	Rajgeera (*Amaranthus paniculatus*)
19	Ajowan	Earthy materials	Exhausted ajowan seeds; excess stem and chaff
20	Mediterra-nean oregano	-	*Origanum majorana; O. syriacum; O. vulgare; Satureja montana*

S. No.	Commodity	Adulterants	
		Non-biological	Biological
21	Asafoetida	Coal tar dyes; gypsum; red clay; chalk	Foreign resin-Gum Arabic, gum resin colophony, galbanum, moriacum, resin, rosin; Barley; wheat or rice flour; slices of potato
22	Saffron	Synthetic dyes-tartrazine, sunset yellow, amaranth, methyl orange, eosin and Erythrosine; honey; glycerine; solutions of potassium or ammonium nitrate sodium sulphate; magnesium sulphate; barium sulphate; borax	Different parts of the saffron flower itself (styles, stamen, strips of the corolla); dried petals of safflower and Scotch marigold; Calendula poppy; arnica; onion skins; stigmata from other species of *Crocus*, pomegranate, Spanish oyster and maize; dyed corn silk; meat fibre; red sandal wood
23	Vanilla beans/extract	Synthetic vanillin, ethyl vanillin, vertraldehyde, piperonal, vanitrope and coumarin	Tonka beans (*Dipteryx odorata*); *Dipteryx oppositifolia*; vanillon (*Vanilla pompona*); little vanilla (*Selenipedium chica*); leaves of orchid *Angreacum fragrans* and *Orchis fusca, spirnathes cernua, Trilisa odoratissima, Nigritella angustifolia and Melilotus* spp.

Intentional adulteration practice is done by the following ways:

a) **Artificially manufactured materials**

In this type of adulteration the original substances are adulterated by the artificially manufactured materials that resemble with original drug. This practice is generally followed for much costlier drugs e.g. nutmeg is adulterated with basswood, paraffin wax is used in place of beeswax, artificial invert sugar is used in place of honey and compressed chicory is used in place of coffee berries.

b) **Inferior quality materials**

In this type, the original drugs are substituted using inferior quality drugs that may be similar in morphological characters but may not have same chemical or therapeutic value as that of the original natural drug. e.g. *Belladonna* leaves are substituted with *Ailanthus* leaves, papaya seeds to adulterate *Piper nigrum*, mangosteen fruits for bael fruits.

c) **Exhausted material**

In this type of substitution, the active constituents of the main drugs are extracted out and are used again by adding colouring or flavouring agents. This technique is frequently adopted for the drugs containing volatile oils, such as clove, fennel, balsam etc.

d) **Foreign matter**

Sometimes synthetic chemicals are used to enhance the natural character e.g. addition of benzyl benzoate to balsam of Peru, Citral to lemon oil etc. Besides the wastes from market are collected and admixed with natural drugs particularly in liquid forms e.g limestone in asafetida, lead shot in opium, white oil in coconut oil.

e) **Powder**

Powdered drugs are found to be adulterated very frequently. Powdered adulterants are generally waste products of suitable colour and density. Examples include powdered olive stones for powdered gentian, liquorice or pepper; brick powder for barks; red sanders wood for chillies; dextrin powder for Ipecacuanha etc.

2. **Indirect or unintentional adulteration**

Sometimes in the absence of proper methods of evaluation, an authentic drug partially or fully devoid of the active ingredients may enter the herbal product market. Some of the major ways in which adulterants get added to the drugs unintentionally are discussed below:

a) **Faulty collection methods**

Some of the herbal adulterations are due to the carelessness of the collectors and suppliers **(Table 9).**

Table 9: Species admixtures in herbal trade

S. No.	Herbal plants/products	Species in admixture	Admixture (%)
01	*Phyllanthus amarus*	*P. debilis, P. fraternus, P. urinaria, P. maderaspatensis* and *P. kozhikodianus*	24
02	*Cassia fistula*	*Senna auriculata*	50
03	*Ruta graveolens*	*Euphorbia dracunculoides*	50
04	*Hypericum perforatum*	*Senna alexandrina*	-
05	*Piper nigrum*	*Capsicum annuum*	22.2
06	*Crocus sativus*	*Arnica montana, Bixa orellana, Calendula officinalis, Carthamus tinctorius, Crocus vernum, Curcuma longa* and *Hemerocallis* species	21
07	*Berberis aristata*	*B. chitria* and *B. lycium*	80
08	*Paris polyphylla*	*Acorus calamus*	-
09	*Citrullus colocynthis*	*Trichosanthes palmata*	-
10	*Swertia chirata*	*Swertia angustifolia*	-
11	*Glycyrrhiza glabra*	*Abrus precatorius*	-
12	*Echinacea purpurea*	*Juglans regia*	-
13	*Bacopa monnieri*	*Eclipta alba* and *Malva rotundifolia*	-

14	*Saraca asoca*	*Polyalthia longifolia, Humboldtia vahliana, Shorea robusta* and *Mallotus nudiflorus*	31
15	*Sida cordifolia*	*Sida acuta*	80
16	*Cinnamomum verum*	*Cinnamomum cassia* and *Cinnamomum malabatrum*	70

Less valuable part of the genuine is collected. Moreover collection should be carried out at a proper season and time, when the active constituents reach maximum. For example *Datura strumarium* leaves should be collected during flowering stage and wild cherry bark in autumn. Besides, collection from other plant by ignorance, due to similarity in the appearance, color, lack of knowledge may lead to adulteration. For example in place of *Aconitum napellus, Aconitum deinorhizum* may be collected or in place of *Rhamnus purshiana* (cascara bark) *Rhamnus colifornica* is generally collected.

b) Name confusions

In different states of the country the same plant is known by different vernacular names, while quite different drugs are known by same name. This creates confusion which is best illustrated by Punarnava and Brahmi. The Indian pharmacopoeia drugs *Trianthema portulacastrum* L. and *Boerhavia diffusa* L. Are both known by the same vernacular name "Punarnava, are therefore wrongly collected.

Sometimes confusion also arises in the name of plant with indigenous systems of medicine, resulting in the faulty collection of material. In ayurveda, 'Parpatta' refers to *Fumaria parviflora* while in siddha, 'Parpadagam' refers to *Mollugo pentaphylla.* Owing to the similarity in the names in traditional systems of medicine, these two herbs are often adulterated. Other such examples include *Casuarina equisetifolia* for *Tamarix indica* and *Aerva lanata* for *Bergenia ciliate.*

c) Lack of knowledge about authentic source

Lack of proper knowledge regarding the authentic source sometimes may also lead to adulteration of drugs. For example the authentic source of Nagakesar ayurvedic drug is *Mesua ferrea.* However, market samples are adulterated with flowers of *Calophyllum inophyllum.* Though the authentic plant is available in plenty throughout the Western Ghats and parts of the Himalayas, but suppliers are unaware of it. Due to such reasons, *C. inophyllum* (which is in the plains) is usually sold as Nagakesar.

d) Imperfect preparation

During collection of herbal material sometimes the associated plant parts also get collected. Non-removal of such undesirable parts or structures results adulteration. For example cork should be removed from ginger rhizome. Further, proper drying conditions should be adhered. Improper drying may lead to unintentional adulteration e.g. if *Digitalis* leaves are dried above 65^{0}C, decomposition of glycosides by enzymatic hydrolysis takes place which eventually reduces the efficacy of the drug.

e) **Incorrect storage**

Faulty storage methods could lead to the loss of the active ingredients and in certain extreme cases, production of toxic metabolites. Inadequate physical factors such as air, humidity, light and temperature can bring about deterioration directly or indirectly. These factors, alone or in combination, can lead to the growth of unwanted organisms such as molds, mites and bacteria in the herbal material. Volatile oils for instance should be protected from light and stored in well closed containers in cool place. *Belladonna* leaves should be stored in moisture free containers; otherwise the enzymatic action may lead to decomposition of its active constituents. Small animals like Mites, nematode worms, insects like moths and beetles can also destroy herbal drugs during storage.

f) **Morphological resemblances**

Due to morphological similarities (appearance, color etc.) the genuine crude drugs are substituted with other plant materials and often sold in the market e.g. *Podophyllum peltatum* is used as a substitute for *P. hexandrum*, *Belladona* leaves are substituted with *Ailanthus* leaves, saffron is admixed with dried flowers of *Carthamus tinctorius*, mother cloves and clove stalks are mixed with clove. Some of the other substituents used in the traded herbal medicinal drugs are enlisted in **Table 10.**

Table 10: Substituents used in some traded medicinal plants

Trade name	Genuine plant	Substituents
Sankhpuspi	*Evolvulus alsinoides*	*Clitorea ternatea; Ephedra gerardiana Convolvulus microphyllus*
Jeevaka	*Malaxis acuminata*	*Tinospora cordifolia*
Ratanjot	*Onosma echinoides*	*Clausena pentaphylla; Canscora decussata*
Brahmi	*Bacopa monnieri*	*Centella asiatica*
Chiraita	*Swertia chirayita*	*Andrographis paniculata*
Ativisha	*Aconitum heterophyllum*	*Cyperus rotundus*
Rishabaka	*Microcystis wallichii*	*Bambusa arundinacea*
Chitraka	*Plumbago zeylanica*	*Baliospermum montanum*
Meda	*Polygonatum cirrihifolium*	*Withania somnifera*
Lakshmana	*Ipomea sepiaria*	*Elephantopus scaber*
Mahameda	*Polygonum verticillatum*	*Hemidesmus indicus*
Pushkarmula	*Iris germanica*	*Saussurea lappa*

Trade name	Genuine plant	Substituents
Kakoli dwaya	*Roscaea procera*	*Asparagus racemosus*
Murva	*Marsedenia tenacissima*	*Rubia cordifolia*
Senna	*Cassia senna*	*Cassia angustifolia*
Kateli	*Solanum xanthocarpam*	*Clerodendron indicum*
Gokhru	*Tribulus terrestris*	*Pedalium murg*
Agnimantha	*Clerodendron phlomoides*	*Premna oblusifolia*
Kutki	*Gentiana kuroo*	*Picrorhiza kurroa*
Rasna	*Pluchea lanceolata*	*Alpinia officinarum; Vanda roxburghii Geranium wallichianum*
Somlata	*Sarcostemma acidum*	*Abies webbiana*
Parpatta	*Fumaria parviflora*	*Mollungo pentaphylla*
Punarnava	*Boerhavia diffusa*	*Trianthema portulacasstrum*

Multiple Choice Questions and Answers

1. Who is known as the father of medicine?
   (a) Aristotle (b) Dioscorides
   (c) Hippocrates (d) Galen
   Ans: c
2. Which among the following plants is known as Indian ginseng?
   (a) *Withania somnifera* (b) *Datura ceratocaula*
   c) *Iberis amara* d) *Capsella bursa pastoris*
   Ans: a
3. Ashwagandha is also known as:
   (a) Shatmuli (b) Sarpgandha
   (c) Withania (d) Foxglove
   Ans: c
4. Vinca alkaloids are isolated from
   (a) *Catharanthus roseus* (b) *Rosco chromogens*
   (c) *Catharanthus indicus* (d) None of these
   Ans: a
5. Vincristine and Vinblastine alkaloids isolated from *Catharanthus roseus* are
   (a) Antidiabetic (b) Anticancerous
   (c) Antifungal (d) Anti-inflammatory
   Ans: b
6. Plant material while processing is dried usually at low temperature in order to:
   (a) Prevent evaporation of compounds
   (b) Prevent degradation of bioactive compounds
   (c) Prevent condensation of molecules
   (d) All of these
   Ans: b
7. Which form of herbal medicine is most commonly practiced in India?
   (a) Ayurvedic (b) Unani
   (c) Acupuncture (d) None of these
   Ans: a

8. Which part of the ginger plant is used for the medicinal purposes?
   (*a*) Flower (*b*) Floral bud
   (*c*) Underground stem (*d*) Leaves
   Ans: c
9. Which country is currently leading for incorporating traditional herbal medicine, acupuncture and modern medicine in its healthcare system?
   (*a*) China (*b*) India
   (*c*) USA (*d*) Germany
   Ans: c
10. Which of the following medicinal plants is known as Gotu kola?
    (*a*) *Withania somnifera* (*b*) *Centella asiatica*
    (*c*) *Catharanthus roseus* (*d*) *Ocimum sanctum*
    Ans: b
11. In Ayurveda Tulsi is known as "the elixir of life" because it is
    (*a*) queen of herbs (*b*) promotes longevity
    (*c*) relieves stress (*d*) All of these
    Ans: b
12. How much of the world's population rely on the herbal medicine?
    (*a*) 70% (*b*) 80%
    (*c*) 90% (*d*) 50%
    Ans: b
13. The national Institute of Ayurveda in India is located at
    (*a*) Chennai (*b*) Kolkatta
    (*c*) Jaipur (*d*) Pune
    Ans: c
14. Extracts of which plant have been used for their immunostimulant properties to fight off viral infections such as colds and influenza?
    (*a*) Indian snakeroot, *Rauwolfia serpentina*
    (*b*) Rosy periwinkle, *Catharanthus roseus*
    (*c*) Valerian, *Valeriana officinalis*
    (*d*) Feverfew, *Tanacetum parthenium*
    e) Purple coneflower, *Echinacea purpurea*
    Ans: e
15. Cyanogenic, steroid, and coumarin are all subcategories of which chemical category of poisonous plants?
    (*a*) Alkaloids
    (*b*) Glycosides
    (*c*) Oxalates
    (*d*) Resins and resinoids
    Ans: b

16. The study of the chemistry and physiologic effects of natural products from plants is referred to as:
    (a) Toxicology
    (b) Pharmacognosy
    (c) Pharmacology
    (d) Pharmacokinetics
    e) Phytochemistry
    Ans: b
17. Which plant-derived alkaloid is used as a treatment for malaria?
    (a) Quinine from *Cinchona* species
    (b) Digoxin from *Digitalis* species
    (c) d-Tubocurarine from *Chondrodendron tomentosum*
    (d) Vincristine from *Catharanthus roseus*
    e) Vinblastine from *Catharanthus roseus*
    Ans: a
18. Plant-derived compounds that have pharmacologic activity at low doses and have a characteristic nitrogen atom in a heterocyclic ring are chemically classified as
    (a) Amino acids
    (b) Glycosides
    (c) Steroids
    (d) Flavonoids
    e) Alkaloids
    Ans: e
19. Drug which is not obtained from seed:
    (a) Nux vomica
    (b) Digitalis
    (c) Stropanthus
    (d) Ispgol
    Ans: b
20. Drug which does not belong to leaves:
    (a) Senna
    (b) Digitalis
    (c) Eucalyptus
    (d) Turmeric
    Ans: d
21. Dried latex of the drug is used, except:
    (a) Opium
    (b) Gatta parcha
    (c) Papain
    (d) Balsam
    Ans: b
22. The roots of following drugs are effective, except:
    (a) Rauwolfia
    (b) Ipecacuanha
    (c) Turmeric
    (d) Aconite
    Ans: d
23. Entire parts of the following drugs are effective, except:
    (a) Ergot
    (b) Belladonna
    (c) Ephedra
    (d) Clove
    Ans: d

24. Select the drug, which is not belonging to glycoside class?
(*a*) Digitalis (*b*) Senna
(*c*) Nux vomica (*d*) Cascara
Ans: b

25. Select the drug, which is not belonging to tannin class?
(*a*) Myrobalam (*d*) Pale catechu
(*c*) Ashoka (*d*) Peppermint
Ans: d

26. Drug not belonging to volatile oil class:
(*a*) Peppermint (*b*) Clove
(*c*) Castor oil (*d*) Garlic
Ans: c

27. Select the drug, which is not showing carminative property?
(*a*) Dill (*b*) Mentha
(*c*) Senna (*d*) Cardamom
Ans: a

28. ______ is not used as expectorant.
(*a*) Ipecacuanha (*b*) Vasaka
(*c*) Liquorice (*d*) Atropine
Ans: d

29. ______ is not used as cardiotonic.
(*a*) Digitalis (*b*) Cinchona
(*c*) Squill (*d*) Stropanthus
Ans: b

30. Drug used as antihypertensive:
(*a*) Rauwolfia (*b*) Digitalis
(*c*) Squill (*d*) Stropanthus
Ans: c

31. _______is used as adrenergic.
(*a*) Catharanthus (*b*) Physostigma
(*c*) Pilocarpus (*d*) Belladonna
Ans: d

32. Drug do not used as anticancer:
(*a*) Podophyllum (*b*) Curare
(*c*) Camptotheca (*d*) Taxus
Ans: b

33. Drug is not used as antirheumatic:
(*a*) Quassia (*b*) Aconite
(*c*) Colchicum (*d*) Guggul
Ans: d

34. Which is used as emetic?
    (*a*) Agar (*b*) Isapghul
    (*c*) Ipecac (*d*) Banana
    Ans: c
35. Drug used as bronchodilator:
    (*a*) Tea (*b*) Liquorice
    (*c*) Ipecacuanha (*d*) Vasaka
    Ans: c
36. Plant used against malaria:
    (*a*) Ashwagandha (*b*) Tulsi
    (*c*) Ginseng (*d*) Artemesia
    Ans: d
37. Alkaloids are ________ type of substances.
    (*a*) Acidic (*b*) Neutral
    (*c*) Chemical (*d*) Basic nitrogenous
    Ans: d
38. Glycosides are condensation products of:
    (*a*) Sugar + aglycone (*b*) Sugar + Protein
    (*c*) Protein + aglycone (*d*) Fats + aglycone
    Ans: a
39. Tannins give which colour with iron compound:
    (*a*) Pale yellow (*b*) Blue black
    (*c*) Light pink (*d*) Orange
    Ans: c
40. Flower bud of which plant have medicinal importance:
    (*a*) Saffron (*b*) Clove
    (*c*) Fig (*d*) Caraway
    Ans: b
41. Which is not an example of organized crude drug:
    (*a*) Digitalis (*b*) Cinchona
    (*c*) Aloe (*d*) Clove
    Ans: a
42. Proto alkaloids are also known as:
    (*a*) Amino alkaloid (*b*) Pseudo alkaloid
    (*c*) True alkaloid (*d*) Extra alkaloid
    Ans: a
43. True alkaloids are always present in plants as:
    (*a*) Salts of inorganic acid (*b*) Salts of organic acid
    (*c*) Salts of meconic acid (*d*) Salts of benzoic acid
    Ans: b

44. An example of pseudoalkaloid is
(*a*) Caffeine (*b*) Hygrine
(*c*) Arecoline (*d*) Lobeline
Ans : a

45. Which drug is not under the chemical class of indole?
(*a*) Quinidine (*b*) Quinine
(*c*) Atropine (*d*) Camptothecin
Ans: c

46. Deadly nightshade is the common name of:
(*a*) Datura (*b*) Belladonna
(*c*) Cinchona (*d*) Coca
Ans: b

47. Which is not the use of argot?
(*a*) Oxytocin (*b*) Prevent post partum hemorrhage
(*c*) Migraine (*d*) Laxative
Ans: d

48. Which is not the action of Stramonium?
(*a*) Anticholinergic (*b*) Mydriatic
(*c*) Antiadrenergic (*d*) Control of Motion sickness
Ans: c

49. Which chemical constituent is not naturally obtained from Opium?
(*a*) Morphine (*b*) Codeine
(*c*) Narcotine (*d*) Heroine
Ans: d

50. 'Thorn apple' is the famous name of:
(*a*) Atropa (*b*) Cinchona
(*c*) Datura (*d*) Ephedra
Ans: c

51. Which of the following does not belong to family Apocyanaceae?
(*a*) Vinca (*b*) Rauwolfia
(*c*) Kurchi (*d*) Gloriosa
Ans: d

52. Which is not the biological source of Cinchona?
(a) Cinchona calisaya *(b) Cinchona officinalis*
(c) Cinchona succirubra *(d) Cinchona indica*
Ans: d

53. Which ergot alkaloid is water-soluble?
(*a*) Ergotamine (*b*) Ergosine
(*c*) Ergocristine (*d*) Ergometrine
Ans: d

54. Ergot contains large number of indole alkaloids, which are the derivatives of:
    (*a*) Lysergic acid (*b*) Benzoic acid
    (*c*) Cinnamic acid (*d*) Meconic acid
    Ans: a
55. Which forms of the ergot alkaloids are especially important?
    (*a*) Laevo (*b*) Dextro
    (*c*) Cis (*d*) Trans
    Ans: a
56. 'Crow fig seed' is the synonym of:
    (*a*) *Strychnous nux vomica* (*b*) *Rauwolfia sepentina*
    (*c*) *Rauwolfia densiflora* (*d*) *Rauwolfia vometoria*
    Ans: a
57. Biological source of nux vomica is:
    (*a*) *Strychnous nux vomica* (*b*) *Strychnous potatorum*
    (*c*) *Strychnous nux blanda* (*d*) *Strychnous wallichiana*
    Ans: a
58. The biologically important chemical constituent of *Rauwolfia* is:
    (*a*) Reserpinc (*b*) Resercinnamine
    (*c*) Ajamalicine (*d*) Desperidine
    Ans: a
59. Which compound is having the most potent analgesic property?
    (*a*) Heroin (*b*) Morphine
    (*c*) Codeine (*d*) Papaverine
    Ans: a
60. Which is not the characteristic feature of an alkaloid?
    (*a*) Molecular structure and nitrogen in the molecule
    (*b*) Basic in nature
    (*c*) Biosynthetically derived
    (*d*) Acidic in nature
    Ans: d
61. Jesuit's bark is the synonym of:
    (*a*) Coca leaves (*b*) Ipecacuanha
    (*c*) Cinchona (*d*) Calumba
    Ans: c
62. Which alkaloid is not present in Cinchona bark?
    (*a*) Quinidine (*b*) Quinine
    (*c*) Quinovin (*d*) Cinchonin
    Ans: c

63. Papaverine, narcotine and narceine are under which chemical class?
(a) Quinoline
(b) Benzyl isoquinoline
(c) Isoquinoline
(d) Benzophenone
Ans: c

64. The most potent pharmacological activity of papaverine is:
(a) Spasmolytic *(b)* Emetic
(c) Antitussive *(d)* Antipyretic
Ans: a

65. Adulterants of *Rauwolfia serpentina* are distinguished on the basis of:
(a) Presence of sclernchyma *(b)* Absence of sclerenchyma
(c) Presence of parenchyma *(d)* Absence of parenchyma
Ans: a

66. Glycoside present in *Stychnous nux vomica* is:
(a) Strychnine *(b)* Brucine
(c) Ajmaline *(d)* Loganine
Ans: d

67. Major pharmacological target of Ephedrine is:
(a) Kidney failure *(b)* Inflammation
(c) Heart failure *(d)* Bronchial asthma
Ans: d

68. The characteristic not associated with alkaloids is:
(a) They all contain nitrogen
(b) Most of non-volatile alkaloids are solid
(c) All the alkaloid contains sulphur
(d) They are physiologically active
Ans: c

69. Thalleoquin test is used for the identification of:
(a) Atropine *(b)* Vincristine
(c) Quinine *(d)* Strychnine
Ans: c

70. Stas-otta process is used for the extraction of:
(a) Glycosides *(b)* Alkaloids
(c) Terpenoids *(d)* Resins
Ans: a

71. Senna mainly contains:
(a) O-glycosides *(b)* N-glycosides
(c) C-glycosides *(d)* S-glycosides
Ans: a

72. Which among the following falls under cyanogenic glycosides?
    (a) Bitter almond *(b)* Black mustard
    (c) Digitalis *(d)* Rhubarb
    Ans: a
73. The drug having cardiotonic activity is:
    (a) Thevetia *(b)* Cochineal
    (c) Rhubarb *(d)* Aloe
    Ans: a
74. Which drug is used as diuretic?
    (a) Quillaia *(b)* Senega
    (c) Ginseng *(d)* Gokhru
    Ans: d
75. The drug used as nerve tonic is:
    (a) Brahmi *(b)* Mordica
    (c) Ginseng *(d)* Senega
    Ans: a
76. Borntrager's test is used for one of the following phytochemicals:
    (a) Anthraquinones *(b)* Flavonoids
    (c) Alkaloids *(d)* Steroids
    Ans: a
77. Saponin glycosides shows one of the following property:
    (a) Laxative *(b)* Anticonvulsant
    (c) Foaming *(d)* Astringent
    Ans: c
78. Among the following, which one is not the example of Cardinolide:
    (a) Digitoxin *(b)* Digoxin
    (c) Gitoxigenin *(d)* Scillarin A
    Ans: d
79. The biological source for Dioscorea is:
    (a) Dioscoria deltoidea *(b) Dioscoria floribunda*
    (c) Dioscoria villosa *(d) Dioscorea compositae*
    Ans: a
80. Yam is the synonym of:
    (a) Stropanthus *(b)* Dioscorea
    (c) Safed musali *(d)* Liquorice
    Ans: b
81. Dioscoria is used in the treatment of:
    (a) Ulcer *(b)* Cancer
    (c) Rheumatic arthritis *(d)* Kidney stone
    Ans: c

82. Liquorice belongs to family:
(*a*) Liliaceae (*b*) Triterpenoid
(*c*) Loganaceae (*d*) Leguminosae
Ans: d

83. Liquorice is traditionally used in the treatment of:
(*a*) Conjuctivitis (*b*) Peptic ulcer
(*c*) Allergy (*d*) Skin disease
Ans: b

84. Rhubarb is cultivated mainly in which state of India:
(*a*) Jammu and Kashmir (*b*) Madhya Pradesh
(*c*) Maharashtra (*d*) Bengal
Ans: a

85. Keller killani test is positive for:
(*a*) Digitoxose (*b*) Gitoxose
(*c*) Digitoxigenin (*d*) Gitoxigenin
Ans: a

86. Presence of star spots is the characteristic of:
(*a*) Cascara (*b*) Senna
(*c*) Aloe (*d*) Rhubarb
Ans: d

87. Botanical source of Arjuna bark is:
(*a*) *Terminalia tamentosa* (*b*) *Polygala chinesis*
(*c*) *Poligala erioptera* (*d*) *Terminalia arjuna*
Ans: d

88. Presence of aleurone grain is the characteristic of:
(*a*) Belladonna (*b*) Nux vomica
(*c*) Hyoscyamus (*d*) Vinca
Ans: b

89. Digitalis leaves should be dried at temperature below:
(*a*) 20 °C (*b*) 30 °C
(*c*) 40 °C (*d*) 60 °C
Ans: d

90. Vein-islet no. of Indian senna is in the range of:
(*a*) 10.5 to 12.5 (*b*) 13.5 to 16.5
(*c*) 16.5 to 18.0 (*d*) 19.5 to 22.5
Ans: d

91. Adulteration by exhausted ginger is detected by its:
(*a*) Shape (*b*) Size
(*c*) Colour (*d*) Water soluble ash
Ans: d

92. Terpenes are made up of:
(*a*) Calcierene (*b*) Pseudopriene
(*c*) Neoprene (*d*) Isoprene
Ans: d

93. Molecular formula for isoprene unit is:
(*a*) $C_8 H_8$ (*b*) $C_3 H_8$
(*c*) $C_5 H_8$ (*d*) $C_6 H_6$
Ans: c

94. Example of alcohol volatile oil is:
(*a*) Peppermint (*b*) Cinnamon
(*c*) Lemon peel (*d*) Orange peel
Ans: a

95. The major terpenoid in clove oil is:
(*a*) Eugenol (*b*) Santalol
(*c*) Anethol (*d*) Menthol
Ans: a

96. Lemon grass oil is obtained from:
(*a*) *Cymbopogon flexuousus* (*b*) *Cinnamonum camphora*
(*c*) *Eucalyptus globules* (*d*) *Chenopodium ambrosioides*
Ans: a

97. Peppermint oil contains chiefly:
(*a*) I-menthol (*b*) d-menthol
(*c*) Citral (*d*) Cineole
Ans: a

98. The chief active constituent of fennel is:
(*a*) Carvone (*b*) Borneol
(*c*) Dill-Apiole (*d*) Fenchone
Ans: d

99. Exhuasted fennel is identified by the absence of:
(*a*) Fenchone (*b*) Anethol
(*c*) Carvone (*d*) Cineol
Ans: a

100. The family of *Ocimum sanctum* is:
(*a*) Labiatae (*b*) Liliaceae
(*c*) Lauraceae (*d*) Loganaceae
Ans: a

101. 'Holy basil' is the synonym for:
(*a*) *Alpinia officinarum* (*b*) *Allium sativum*
(*c*) *Nardostachys jatamonum* (*d*) *Ocimum sanctum*
Ans: d

102. Alkaloid present in black pepper is:
(a) Caryophyllene (b) Cinnamaldehyde
(c) 1-phellandrene (d) Piperine
Ans: d

103. Goldbeater skin test is used for the identification of:
(a) Alkaloids (b) Glycosides
(c) Tannins (c) Resins
Ans: c

104. Condensed tannins are also called:
(a) Hydrolysable tannins (b) Non-hydrolysable tannins
(c) Pseudotannins (d) Prototannins
Ans: b

105. Chemical test used for the identification of carbohydrates is:
(a) Ninhydrin (b) Molisch
(c) Borntrager (d) Baljet
Ans: b

106. Shatavari is under which class of glycoside:
(a) Sterol (b) Saponin
(c) Cyanogentic (d) Isothiocyanate
Ans: b

107. The medical system developed by ancient sages in Tamil Nadu is called:
(a) Unani (b) Siddha
(c) Ayurvedic (d) Western
Ans: b

108. A substance that promotes emptying of the bowels is:
(a) Lactifuge (b) Nervine
(c) Laxative (d) Tonic
Ans: c

109. Which among the following is also called Indian gooseberry?
(a) Trigonella foenum graceum
(b) Phyllanthus emblica
(c) Centella asiatica
(d) Dioscorea deltoidea
Ans: b

110. Salkowski test is used for the identification of:
(a) Steroids
(b) Alkaloids
(c) Glycosides
(d) Triterpenoids
Ans: a

Choose the Correct Pairs

1. **Part used** — **Plant**

Part used	Plant
(a) Root	i) Quassia
(b) Wood	ii) Turmeric
(c) Rhizome	iii) Karachi
(d) Bark	iv) Rauwolfia

Ans: a) iv b) i c) ii d) iii

2. **Part used** — **Example**

Part used	Example
(a) Seeds	i) Vasaks
(b) Leaves	ii) Coriander
(c) Fruits	iii) Pyrethrum
(d) Flowers	iv) Isapgol

Ans: a) iv b) i c) ii d) iii

3. **Phytocompound class** — **Example**

Phytocompound class	Example
(a) Glycosides	i) Cinchona
(b) Alkaloids	ii) Digitalis
(c) Tannins	iii) Ashoka
(d) Carbohydrates	iv) Guar gum

Ans: a) ii b) i c) iii d) iv

4. **Medicinal use** — **Example**

Medicinal use	Example
(a) Emetic	i) Castor oil
(b) Purgative	ii) Cardamom
(c) Carminative	iii) Cinchona
(d) Bitter	iv) Ipecacuanha

Ans: a) iv b) i c) ii d) iii

5. **Pharmacological action** — **Example**

Pharmacological action	Example
(a) Anticholinergic	i) Physostigmina
(b) Cholinergic	ii) Datura
(c) Adrenergic	iii) Opium
(d) Expectorant	iv) Vasaka

Ans: a) ii b) I c) iii d) iv

	Reagent	Composition
6.	*(a)* Mayer's	i) Potassium bismuth iodide
	(b) Dragendorff's	ii) Iodine in potassium iodide
	(c) Wager reagent	iii) Picric acid
	(d) Hager's reagent	iv) Potassium mercuric iodide

Ans: a) iv b) i c) ii d) iii

	Alkaloid	Family
7.	*(a)* Ergot	i) Papaveraceae
	(b) Nux vomica	ii) Clavicipitaceae
	(c) Opium	iii) Loganiaceae

Ans: a) ii b) iii c) i

	Drug	Activity
8.	*(a)* Ergot	i) CNS stimulant
	(b) Nux vomica	ii) Antimitotic
	(c) Rauwolfia	iii) Oxytoxic
	(d) Vinca	iv) Hypotensive

Ans: a) iii b) i c) iv d) ii

	Drug	Active constituent
9.	*(a)* Ergot	i) Papaverine
	(b) Opium	ii) d-tubocurarine
	(c) Curare	iii) Cephaeline
	(d) Ipecac	iv) Ergotamine

Ans: a) iv b) i c) ii d) iii

	Drug	Synonym
10.	*(a)* Nux vomica	i) Chota chand
	(b) Rauwolfia	ii) Periwinkle
	(c) Vinca	iii) Strychni Semen
	(d) Hyoscyamus	iv) Henbane

Ans: a) iii b) i c) ii d) iv

	Drug	Family
11.	*(a)* Indian senna	i) Liliaceae
	(b) Aloe	ii) Polygonaceae
	(c) Rhubarb	iii) Rhamnaceae
	(d) Cascara	iv) Leguminoseae

Ans: a) iv b) i c) ii d) iii

	Glycosides	Example
12.	*(a)* Anthracene glycoside	i) Digitalis
	(b) Cardiac glycoside	ii) Bitter almond
	(c) Saponin glycoside	iii) Dioscorea
	(d) Cyanogenetic glycoside	iv) Senna leaf

Ans: a) iv b) i c) iii d) ii

13. **Drug** | **Activity**
(*a*) Red squill | i) Cardiotonic
(*b*) Kalmegh | ii) Purgative
(*c*) Aloe | iii) Nervine tonic
(*d*) Brahmi | iv) Rat poisoning
Ans: a) iv b) i c) ii d) iii

14. **Drug** | **Use**
(*a*) Gokhru | i) Hypoglycemic
(*b*) Momordica | ii) Galactogogue
(*c*) Shatavari | iii) Diuretic
(*d*) Liquorice | iv) Expectorant
Ans: a) iii b) i c) ii d) iv

15. **Drug** | **Source**
(*a*) Gokhru | i) *Glycyrrhiza glabra*
(*b*) Liquorice | ii) *Urgenia martima*
(*c*) Bramhi | iii) *Tribulus terrestris*
(*d*) European squill | iv) *Bacopa moniera*
Ans: a) iii b) i c) iv d) ii

16. **Drug** | **Test**
(*a*) Aloe | i) Borntrager test
(*b*) Senna | ii) Baljet test
(*c*) Digitalis | iii) Killer killani test
(*d*) Deoxy sugar | iv) Modified anthraquinones test
Ans: a) iv b) i c) ii d) iii

17. **Drug** | **Part used**
(*a*) Aloe | i) Leaf
(*b*) Digitalis | ii) Bulb
(*c*) Indian squill | iii) Seed
(*d*) Stropanthus | iv) Dried juice
Ans: a) iv b) i c) ii d) iii

18. **Drug** | **Source**
(*a*) Seame oil | i) *Hydrocarpus wightiana*
(*b*) Chaulmoogra oil | ii) *Zea mays*
(*c*) Corn oil | iii) *Sesamum indicum*
(*d*) Safflower oil | iv) *Carthmus tinctorius*
Ans: a) iii b) i c) ii d) iv

19. **Volatile oil** | **Source**
(*a*) Turpentine oil | i) *Cymbopogon citratus*
(*b*) Lemon grass oil | ii) *Santalum album*
(*c*) Sandal wood oil | iii) *Artemisia pallens*
(*d*) Davana oil | iv) *Pinus species*
Ans: a) iv b) i c) ii d) iii

20.

Drug	**Source**
(a) Cardamom	i) *Coriandrum sativum*
(b) Caraway	ii) *Carum carvi*
(c) Coriander	iii) *Anethum graveolens*
(d) Dill	iv) *Elettaria cardamomum*

Ans: a) iv b) ii c) i d) iii

21.

Adulterants for clove	**Characteristics**
(a) Mother clove	i) Expanded flowers of clove trees
(b) Blown cloves	ii) Contain only 5% of oil
(c) Clove stalks	iii) Oil is removed from clove
(d) Exhausted clove	iv) Dark brown, ovate and ripened fruits of clove

Ans: a) iv b) i c) ii d) iii

22.

Medicinal plant	**Family**
(a) Zingiber offcinale	i) Solanaceae
(b) Capsicum annum	ii) Zingiberaceae
(c) Cannabis sativa	iii) Convolvulaceae
(d) Ipomoea purga	iv) Cannabaceae
(e) Podophyllum hexandrum	v) Berberidaceae

Ans: a) ii b) i c) iv d) iii e) v

23.

Drug	**Phytochemical**
(a) Ginger	i) Capsaicin
(b) Capsicum	ii) Alpha elaterin
(c) Colocynth	iii) Benzoic acid
(d) Benzoin	iv) Zingerone

Ans: a) iv b) i c) ii d) iii

24.

Drug	**Alternative name**
(a) Turmeric	i) Indian saffrom
(b) Capsicum	ii) Devil's dung
(c) Asafoetida	iii) Maxican jalap
(d) Jalap	iv) Cayenne pepper

Ans: a) i b) iv c) ii d) iii

Appendix

Glossary of Medical Terms

Abatement	A decrease or easing of pain or any other symptom or sign
Abduction	The movement of a limb away from the midline of the body
Abortion	Premature exit of the products of conception (the fetus, fetal membranes and placenta) from the uterus
Acid reflux	A common and abnormal condition in which acid in the stomach rises up into the esophagus
Acupuncture	The practice of inserting needles into the body to reduce pain or induce anesthesia
Aging	The process of becoming older, a process that is genetically determined and environmentally modulated
AIDS	Acquired Immunodeficiency Syndrome, a syndrome caused by infection with the HIV, with ensuing compromise of the body's immune system
Allopathic	Conventional medicine, the usual practice of medicine as opposed to homeopathy
Amnesia	An impairment to or lack of memory
Anemia	The condition of having a lower-than-normal number of red blood cells or quantity of hemoglobin
Anesthesia	Loss of feeling or awareness, as when an anesthetic is administered before surgery
Angiography	A procedure performed to view blood vessels after injecting them with a radiopaque dye that outlines them on x-ray
Antidote	A drug that counteracts a poison
Appendicitis	Inflammation of the appendix, associated with infection of the appendix
Beriberi	Inflammation of multiple nerves due to a deficiency of thiamine (vitamin B1) in the diet

Bile	A yellow-green fluid made by liver, stored in the gallbladder and helps in the digestion of fat
Bronchitis	Inflammation and swelling of the bronchi
Calcitonin	Ahormone produced by thyroid gland that lowers the levels of calcium and phosphate in the blood and promotes the formation of bone
Carcinogen	A substance or an agent that causes cancer
Clinical trial	Study that is intended to evaluate the safety and effectiveness of medications by monitoring their effects on large groups of people
Dengue	An acute mosquito-borne viral illness of sudden onset, characterised by headache, fever, joint pain, muscle pain, swollen glands and rash
Drug resistance	The ability of microorganisms to withstand a drug that once stalled them or killed them
DOTS	Stands for directly observed treatment, a strategy used to reduce the number of tuberculosis cases
Dysuria	Pain during urination, caused by inflammation of urethra, as a result of infection
ECG	Electrocardiogram (used to reflect underlying heart conditions by measuring the electrical activity of the heart
Eczema	Also called atopic dermatitis, is a type of inflammatory reaction of the skin with itching and burning
Edema	Swelling of soft tissues as a result of excess fluid accumulation
Expectorant	A medication that helps bring up mucus and other material from the lungs, bronchi and trachea
Fecundity	The ability to have lots of offsprings with ease
Femicide	The killing of women
Fungicide	Any agent capable of killing, controling or destroying fungi
Gallbladder	A pear shaped organ located below the liver that stores the bile secreted by the liver
Gametogenesis	The development and production of male and female germ cells
Gastritis	Inflammation of the stomach
Gonorrhea	An infection that is transmitted by sexual contact and caused by *Neisseria gonorrhoeae*

Hallucination	Sensory experiences in which a person can see, hear, smell, taste or feel something that is not there
Hay fever	A seasonal allergy to airborne particles characterised by itchy eyes, runny nose, nasal congestion, sneezing and excess mucus
Heartburn	An uncomfortable feeling of burning and warmth occuring in waves rising up behind the breastbone towards the neck
Hematopoiesis	The production, development and differentiation of blood cells from stem cells
IBS	Irritable bowel syndrome (a common gastrointestinal disorder involving an abnormal condition of gut motility and sensation)
ICU	Intensive care unit (the designated area of the hospital that is dedicated to the care of patients who are seriously ill)
Illusion	A perception that occurs when a sensory stimulus is present but is incorrectly perceived and misinterpreted
Immunity	The condition of being immune
In vitro	A test that is done in glass or plastic vessels in the laboratory
Infertility	Diminished or lack of ability to conceive and bear offspring
Jaundice	Yellow staining of the skin and sclerae by abnormally high blood levels of the bilirubin
Juvenile	Between infantile and adult
Kala-azar	A chronic and potentially fatal disease of the viscera, particularly of liver, spleen, bone marrow and lymph nodes due to *Leishmania donovani*
Karyotype	A standard arrangement of the chromosome complement prepared for chromosome analysis
Kinship	Relationship by marriage or, specially, a blood tie
Lactation	The process of milk production by mammary glands
Laparoscopy	A type of surgery in which small incisions are made in the abdominal wall through which a laparoscope and other instruments can be placed to permit structures within the abdomen and pelvis to be seen
Laxative	Something that promotes emptying of the bowels
Leukemia	Cancer of the blood cells
Malformation	A structural defect in the body due to abnormal embryonic or fetal development
Meditation	A typically self directed practice for relaxing the body and calming the mind

Menopause	The time in a woman's life when menstrual periods permanently stop
Migraine	Periodic attacks of headaches on one or both sides of the head
Naturopathy	A system of therapy that is based on preventive care and on the use of physical forces as primary therapies for disease
Nebulizer	A device for administering a medication by spraying a fine mist
NSAID	Non-steroidal anti-inflammatory drug, to treat the inflammatory conditions associated with different disorders
Obesity	The state of being well above one's normal weight
Oncology	The field of medicine that is devoted to cancer
ORS	Oral rehydration solution
Pacemaker	A system that sends electrical impulses to the heart in order to set the heart rhythm
Paralysis	Loss of voluntary movement (motor function)
Pathologist	A physician who identifies diseases and conditions by studying abnormal cells and tissues
Pneumonia	Inflammation of lungs by the infection of bacteria, virus, fungus or other parasite
Puberty	The period in which the human body first becomes capable of reproduction
Q fever	An infectious disease caused by *Coxiella burnetii* bacterium
Quarantine	A period of isolation decreed to control the spread of infectious disease
Radiologist	A physician specialized in radiology, branch of medicine that uses ionizing and non-ionizing radiations for the diagnosis and treatment
Rheumatism	Painful conditions of muscles, tendons, joints and bones
Rh factor	An antigen found on the surface of red blood cells
Rickets	A disease of infants and children that disturbs normal bone formation, leading to failure to mineralize bone
Sarcoma	One of a group of tumors arising from connective tissue
Sedative	A drug that calms a patient, easing agitation and permitting sleep
Seizure	Uncontrolled electrical activity in the brain, which may produce a physical convulsion

Spondylitis	Inflammation of one or more of the vertebrae o f the spine
Swine flu	A respiratory illness caused by infection with swine influenza A virus
Tuberculosis	A highly contagious infection caused by *Mycobacterium tuberculosis*
Turner syndrome	The most common sex disorder in females, characterized by short stature, webbed neck, broad shield-like chest, wide spaced nipples, short fourth finger etc.
Ultrasound	High-frequency sound waves, used to produce a picture of tissues called sonogram
Urolithiasis	The process of forming stones in the kidney, bladder and or urethra
Vaccination	Injection of a killed microbe in order to stimulate the immune system against the microbe
Ventilator	A machine that mechanically assist a patient in the exchange of oxygen and carbon dioxide
Virulent	Extremely noxious, damaging, deleterious and disease causing, marked by a rapid, severe and malignant course
Viscera	The internal organs of the body, especially within the chest or abdomen
Wart	A local growth of the outer layer of the skin caused by a virus
Whooping cough	A highly contagious acute respiratory illness caused by *Bordetella pertussis* bacteria, characterized by whooping sound while inhaling
Xenobiotic	Natural substances that is foreign to the body
Xeroderma	Abnormally dry skin, caused by vitamin A deficiency, systemic illness, overexposure to sunlight or medication
YAC	Yeast artificial chromosome, a vector used in the laboratory to clone pieces of DNA
Yellow fever	An acute systemic illness caused by a virus, *Flavivirus*, characterized by yellowing of skin
Zonules	Thin, gelatinous ligaments that attach the lens to the ciliary body and support the lens centrally behind the pupil
Zoonosis	An infection that is known in nature to infect both humans and lower vertebrate animals